Ho Math, Chess, and Puzzles
For Grade 1 and Under

低年级棋谜式数学

何算独棋
何谜宫棋
智力数谜

Frank Ho   Amanda Ho

何数棋谜 培训

Ho Math Chess Learning Centre

More info on Ho Math Chess and its worksheets, watch the following video.

Student's Name _____ Date _____

Table of Contents

Ho Math Chess

何数棋谜

Preface

I started to get involved in chess teaching at the time when my son was about 6 years old. He showed interest in a set of chess placed on the table in the living room. He was excited about finding out that by making a right move, he was rewarded with taking his opponent's piece. Later we studied chess together, and he became a Canadian junior chess champion and a FIDE chess master.

My interest shifted from playing chess itself to studying the relation between chess and mathematics. Eventually, I created this world's first *Ho Math, Chess, and Puzzles for Grade 1 and Under* workbook. There are many pure math puzzles or chess puzzles only workbooks on the market. How this *Ho Math, Chess, and Puzzles for Grade 1 and Under* workbook sets itself apart from others is that it contains one-of-a-kind mathematical chess puzzles in which math concepts and chess knowledge are integrated using my invention of SCL (Symbolic Chess Language) to link math and chess, and this discovery is the world's first. SCL is based on the chess set, which I created using a geometry concept of line segments and lines.

This workbook is a revolutionary breakthrough in teaching math using a game-based teaching method. Many problems require children to look for problems using spatial relations and patterns, also find their answers by computing in multi-step. As a result, this math and chess integrated workbook takes the boredom and repetitions away from computing practice. Ho Math Chess teaching method can improve children's math scores and problem-solving ability; develop children's logic and critical thinking skills. It is natural health food for improving brainpower.

Most of the teaching organizations use a very classic teaching method that is to teach basic numeracy or counting skills, and they do not know how to integrate a chess game into math worksheets systematically. Ho Math Chess is the only one and the world's first to have integrated chess and puzzles into math calculation worksheets to make math learning fun.

This workbook uses chess pieces, chess points and chessboards etc. to instil mathematical concepts related to the pattern, logic, geometry transformation, number theory and many, many other math concepts.

This workbook is created with the ideas of learning math in multi-concept, multi-sensory, multi-direction and hands-on approach. It is an innovative product of supplementing elementary mathematics. The purposes of this workbook are to develop and train the skills of the following areas.

Student's Name _____   Date _____

- Building basic facts skills

  Our invention of proprietary SCL (Symbolic Chess Language) provides many multi-step problems to train computation skills.

- Transferring abstract idea to concrete numeric values

  SCL provides training to the understanding of how symbols are converted to meaningful numerical values.

- Developing reasoning and logic skills

  Math and Chess integrated teaching method is one of the best tools to train logic and critical thinking skills.

- Processing image and data analysis

  The math and chess integrated problem provides ample opportunities to process chess diagrams, to analyze data, to compare and then to conclude.

- Training visualization and spatial reasoning

  Math and chess integrated problems provide training in visualization, number relations, and spatial reasoning.

- Developing hand-eye-brain coordination with hands-on experience

  Chess provides the best training tool for hand-eye-brain coordination to develop cognitive ability.

Frank Ho
Amanda Ho

November 2017

This workbook can be used to teach kindergarten or grade 1 basic computations, including addition and subtraction.

December 2019

Frank

Student's Name＿＿＿＿＿＿＿＿＿＿＿＿　　　　　　　Date＿＿＿＿＿＿＿

Ho Math Chess workbooks are good for children. (何数棋谜教材对兒童的好処)

加拿大　何数棋谜　培训中心
創辦人何数棋（Frank Ho）

今天儿童面对的世界是学习如何处理数字,图形,资料搜寻,音影上下载,资讯比较,分类等资讯. 这些活动实际已成為儿童生活的一部份.所以如果说学数学就是计算数字就错了.学数学的另 一个目的就是学习如何利用数字资讯去解决问题及培养创造力.但是**传统式数学的计算练习 题却完全没跟上科研已经改变了儿童面对的世界.**

儿童想要的计算题已经不是单纯的从上到下,从左到右的纯计算.**儿童需要的是他们情愿的 而又快乐地做不枯燥的计算题.**所以如何将传统式数学计算题变得有趣而且又好玩,并且还 可以增强儿童的计算及解决问题能力及培养创造力,同时还可以增进儿童记忆的能力达到全 脑开发的目的?

何数棋谜首创已申情商杯的几何棋艺符号并利用此符号发明了世界第一無二的何数棋述教 材及教学棋具.何数棋谜教材让儿童能利用几何棋艺符号进行数学的运算.

何数棋谜与传统式,数学教材不同的是小朋友不但要发掘题目,而且还要依国际象棋棋子的 走法去发掘谜题与计算题（見下图）及答案. **只见棋谜不见题　劝君述路不哭涕　数学 象棋加谜题　健脑思维真神奇**

何数棋谜是将国际象棋融入数学以达到寓教於乐的教学理念.学生不但可以增强计算能力 并且还可以增强解题能力及培养全脑开发创造力.

详细资料请上网查询 **www.homathchess.com.**

Student's Name _____ Date _____

Ho Math Chess worksheet simulating an internet screen (何数棋谜教材与互联网紧密结合**)**

Ho Math Chess worksheet layout simulates
a computer screen and a cell phone screen.

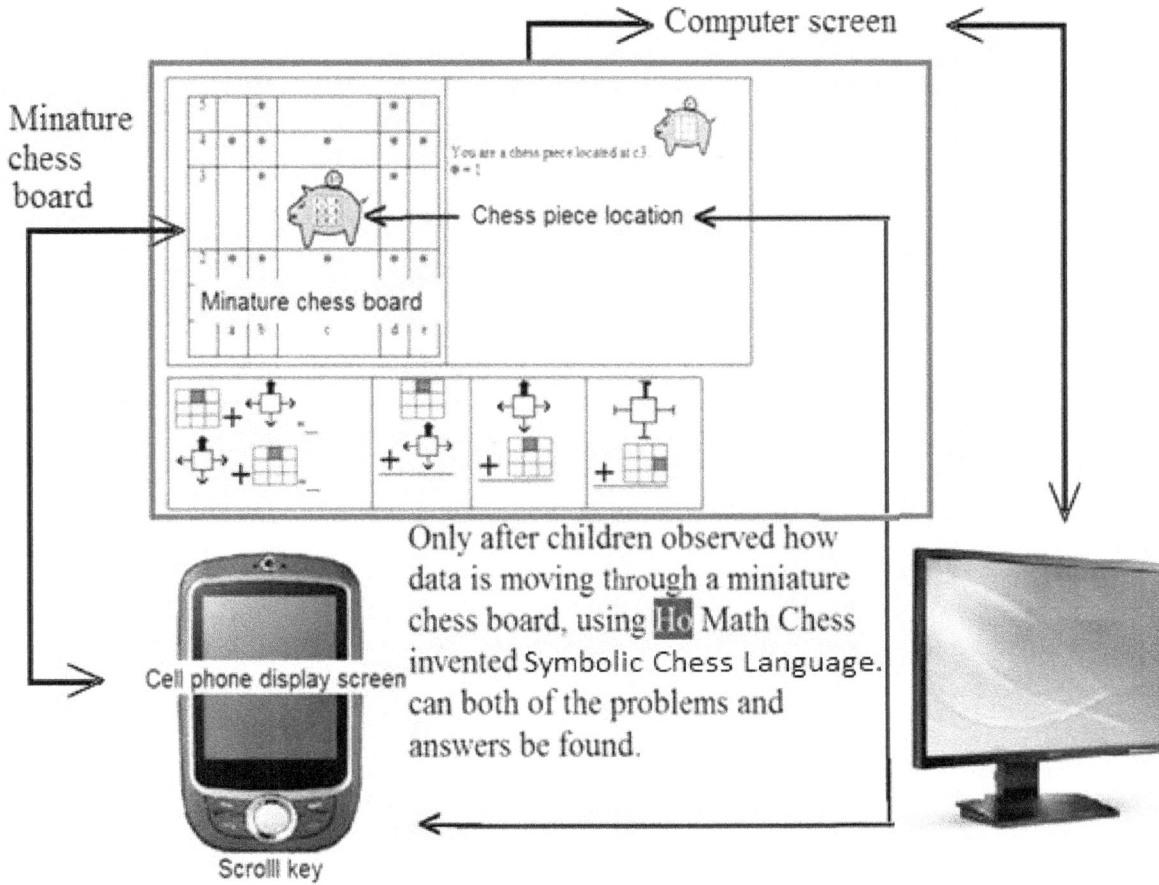

Computer screen

Minature chess board

Minature chess board

You are a chess piece located at c3.
※ = 1

Chess piece location

Cell phone display screen

Scrolll key

Only after children observed how
data is moving through a miniature
chess board, using Ho Math Chess
invented Symbolic Chess Language,
can both of the problems and
answers be found.

Math and chess integrated puzzles

Symbolic Chess Language (SCL) for a flat chess set

Frank Ho, the founder of Ho Math Chess Learning Centre headquartered in Vancouver, Canada, invented a chess language called Symbolic Chess Language (SCL) to link math, chess and puzzles. Each chess symbol represents a corresponding physical chess piece (Figures 1 and 2). This set of chess symbols not only makes the teaching chess easier for younger children as young as 4 years old, but it also serves as a set of command language to link arithmetic and chess. The teaching idea of using this set of chess symbols is to create math and chess integrated problems or any variations of future problems as results of using these symbols. This set of chess symbols and their teaching method have been approved for intellectual property international copyright protection. Problems shown herein are merely exemplary and may be changed to suit different types of problems. Accordingly, the inventor intends to embrace all such alternatives, modifications, and variations as fall within the spirit and broad scope of this invention.

As far as the training of playing chess itself is concerned, SCL symbols have many advantages over the regular chess fonts or traditional 3 D figurines in training children's critical thinking skills when integrating chess into math. The transformation concept in geometric chess symbols is self-explanatory and it is easier for children to understand when each symbol direction pointed by an arrow representing the actual movement direction of each chess piece. Children get hands-on experience in moving those pieces by simply following the directions displayed on each chess piece. The arrow represents a line of math, which means the chess piece can move if it is safe to do so. The line segment means the chess piece can only make one move.

The other advantage of using SCL is that they provide children with opportunities in learning important math concepts in patterns, sequence, symmetry, and transformation related math problems. For example, a typical problem might involve how a 3D object such as a chess piece is transformed into a symbol (or) and then a symbol (or) is translated into a number (9), and finally, a numerical value is produced as an answer.

A mathematical symbol language using the set of SCL can be developed to create an array of innovative arithmetic problems since these geometric chess symbols themselves representing the moving direction of chess figurines. For example, a black rook is represented by this symbol (or) and a highlighted arrow such as indicating its direction of movement towards the right. In this rook case, the symbol not only can represent the chess piece itself, but it also has another attribute which has the 4 directions (up, down, left and right) of moving. The directions can be one way, two ways, three ways, or four ways, so altogether, there could be 15 ways of moving directions. A simple rook's move problem could become a very challenging problem when combined with arithmetic computation problems.

The effect is children feel thrilled and are more willing to work on chess and math combined problem since each problem requires children's creativity to create the questions by following a puzzle-like mini question and the requirement of having children to write the questions reinforces the task of memorizing the basics facts of addition, subtraction, or multiplication without causing stress on children.

Ho Math Chess™ believes the invention of this SCL has brought integrated math and chess teaching to a new horizon, and we are very proud to be the leader in the continued research of math and chess integrated teaching.

Figure 1 SCL (Symbolic Chess Language) for black pieces

Points	1	5	3	3	0	9
Symbols of 3 D traditional chess pieces						
English name	Pawn	Rook	Knight	Bishop	King	Queen
Symbols of flat chess pieces						
SCL handwriting symbol						

Figure 2 SCL (Symbolic Chess Language) for White pieces

Points	1	5	3	3	0	9
Symbols of 3 D traditional chess pieces						
English name	Pawn	Rook	Knight	Bishop	King	Queen
Symbols of flat chess pieces						
SCL handwriting symbol						

The black and white colours of chess symbols along with their black and white squares and their different font sizes present the possibilities of creating a variety of pattern problems.

Setup of chessboard using flat chess set symbols

The setup of a chessboard using real flat chess pieces is shown on the left, and their matching symbols (SCL) are shown on the right.

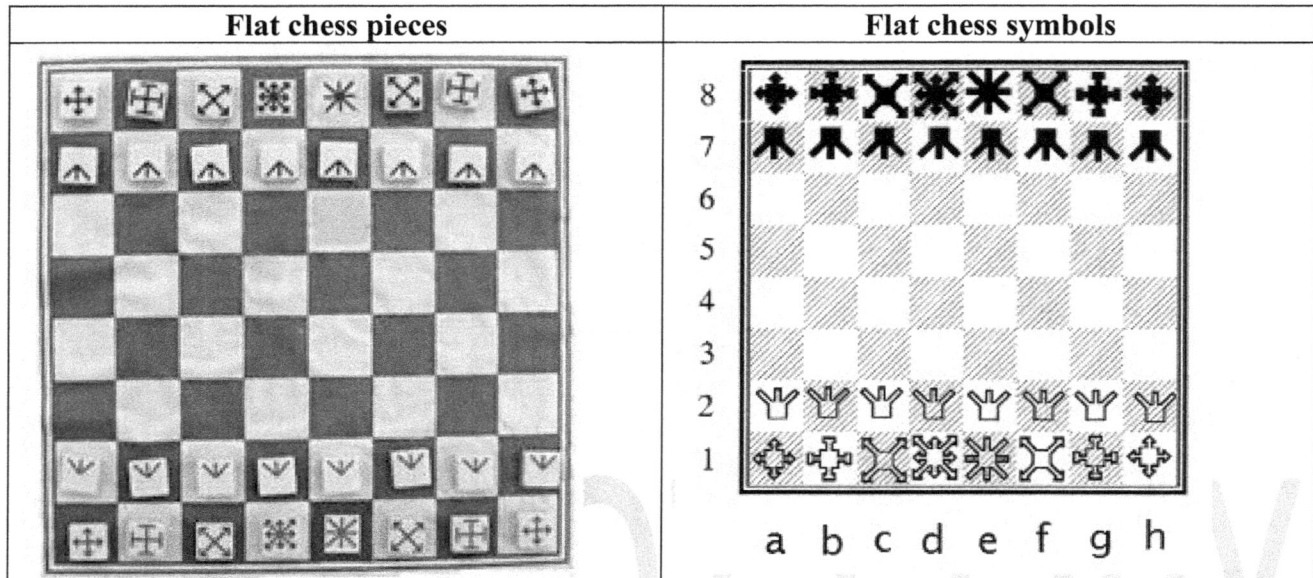

Flat chess pieces	Flat chess symbols

Teaching ideas.

Mention to children that the right-hand side cornet must be a white square for both players, and there are numbers from top to down on the side and letters at the bottom from left to right.

The purpose of letters and numbers but to give each square a name. Explain what a square is?
Ask children to count out loud the rows from top to down and columns from left to the right.

Explain how a chess game is played with 2 players and one side has light pieces, and the other has dark pieces.

Student's Name _____ Date _____

Sorting chess symbols attributes

If students understand the colours of chess pieces and the colours of their occupied squares, it will also help them understand how the chess piece moves. For example, a knight always moves to a different colour square. A black-square-bishop, the bishop sitting on a black square, always moves along black squares, so it can never take or protect any pieces sitting on the white squares. The queen or rook can travel either black or white square but must follow its directions. The following is an example of different chess fonts with different colours of squares.

Replace each ? by the number of the count of each chess piece of the above chess diagram.

1	?	?	?	?	?	?	?

?	?	?	?	?	?	?	?

?	?	?	?	?	?	?	?

?	?	?	?	?	?	?	?

Student's Name _____   Date _____

Chess pieces and their mathematical values

Symbols of chess pieces The left 2 are flat chess symbols (SCL). The right 2 are 3 D chess symbols.	Names of chess pieces	Mathematical values
	Queen	9
	Rook	5
	Bishop	3
	Knight	3
	Pawn	1
	King	0

Student's Name ______________________ Date ______________

Matching SCL flat chess symbol and their corresponding 3D chess pieces

Symbolic Chess Symbols (SCL)	Draw lines	3D chess symbols
		♔
		♘
		♙
		♕
		♖
		♗
		♜
		♝
		♛
		♟
		♞
		♚

Student's Name _____   Date _____

Matching column 1 to column 2 with the equivalent chess symbols and occupying the square

Student's Name _____ Date _____

Circling all chess pieces symbols and chess pieces matching each chess value

Chess symbols	Chess values	Answer
	3	
	9	
	3	
	3	
	1	
	5	
	0	
	3	
	5	
	3	
	9	

Student's Name _____　　Date _____

Counting 3D chess symbols and flat chess symbols

Chess pieces or Chess symbols	Counting+
	How many rooks? _____
	How many bishops? _____
	How many queens? _____
	How many knights? _____
	How many pawns? _____
	How many kings? _____

Student's Name _____   Date _____

Matching chess symbols to coin values

Coins	Value of chess symbol(s)
1 cent A.	1. ⬌
5 cents B.	2. ♙
10 cents C.	3. ⬌ + ⬌ + ⬌ + ⬌ + ⬌
25 cents D.	4. ⬌ + ♖

Student's Name _____   Date _____

Draw line to match the value of each coin to the total value of the chess symbol(s).

1 cent	5 cents	5 cents	10 cents	1 cent
5 cents	1 cent	10 cents	1 cent	10 cents
1 cent	10 cents	♛	10 cents	5 cents
10 cents	10 cents	1 cent	5 cents	10 cents
5 cents	1 cent	5 cents	10 cents	1 cent

	Number of 1 cent is	Number of 5 cents is	Number of 10 cents is
	_____	_____	_____
	Number of 1 cent is	Number of 5 cents is	Number of 10 cents is
	_____	_____	_____
	Number of 1 cent is	Number of 5 cents is	Number of 10 cents is
	_____	_____	_____

Student's Name _____   Date _____

Matching shapes

Draw lines to match left side to the right side

Chess pieces and math

1. Find all the chess pieces combinations of chess pieces that can make the same value as a rook value. One is done for you.

$\Downarrow$	✚	✳	✛
0	0	0	1

2. Pattern.

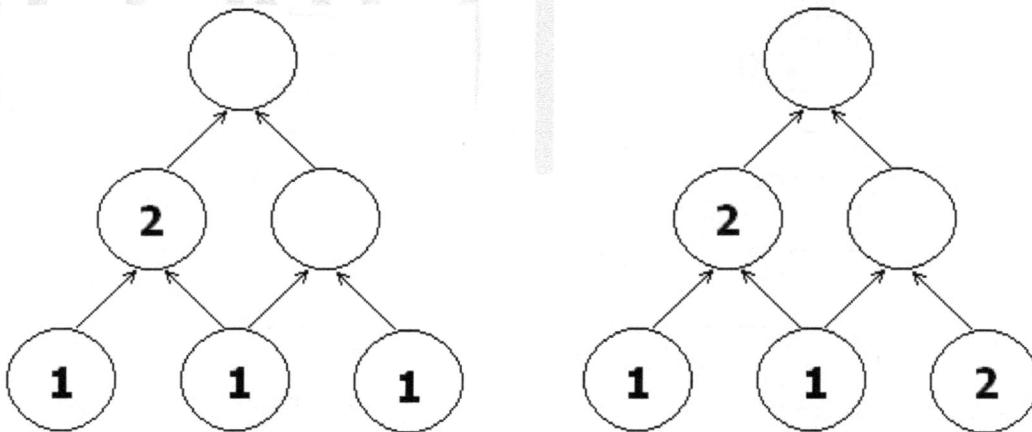

3. At the end of a chess game, White had 3 points left. What possible pieces White could have? (Hint: three possibilities.)

Student's Name _____   Date _____

Using SCL as a tool to link math and chess.

When SCL is used in combination with a chessboard, an array of interesting problems can be created. SCL can be used as a chess notation to record chess moves, and they can also be used as a math operator to do arithmetic computations.

The following demonstrates the concept described above. **Skip this sheet if the students are not ready to do computations.**

	Calculating follows directions
3 4 5 7 9 1 6 4 ♛ 4 6 5 7 6 8 4 3	 Calculate the result of only the numbered squares as directed above _____ .

More complicated problems can be created by combining the direction with coordinates specified. The following example demonstrates the idea described above.

Move ♖ consecutively according to the instructions below.

⬌ e4, 2 (move to right 2 squares) is

____ .

⬍ , 3 is ____ .

⬌ , 4 is ____ .

⬍ , 5 is ____ .

Using SCL as a tool to record chess moves

A reverse problem requires the calculation of moves in backwards. The following example demonstrates the idea.

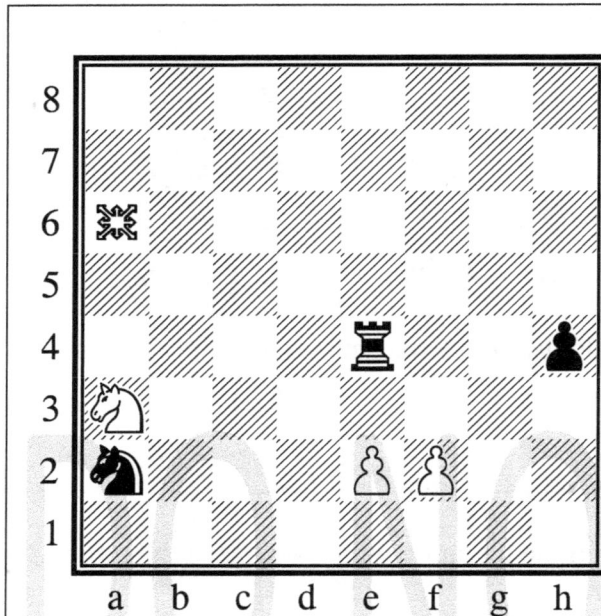

Move ♖ according to the instructions below.

♔e4 to ♔g4 will require ♖ to move ___ squares.

♔ e4 to ♔b4 will require ♖ to move ___ squares.

♔e4 to ♔e7 will require ♖ to move ___ squares.

Student's Name _____ Date _____

Using SCL to record chess moves

Move Rc2 to Rf5 in the shortest moves by highlighting the direction on each of the following ✛.

Path 1: Rc2 ⬌⬍ ⬌⬍ Rf2

Path 2: Rc2 ⬌⬍ ⬌⬍ Rf2

Move Bc2 to Bd5 in the shortest moves by highlighting the direction on each of the following ⤬.

Path 1: Bc2 ⤬ ⤬ Bd5

Path 2: Bc2 ⤬ ⤬ Bd5

Using SCL to record chess moves

Move Nc2 to Nf5 in the shortest moves by highlighting the direction on each of the following ┼ .

Path 1: Nc2 ┼ ┼ Nf5

Path 1: Nc2 ┼ ┼ Nf5

Move Nc2 to Nd5 in the shortest moves by highlighting the direction on each of the following ┼ .

Path 1: Nc2 ┼ ┼ Nf5

Path 1: Nc2 ┼ ┼ Nf5

Connecting the dots according to the number sequence starting from 1.

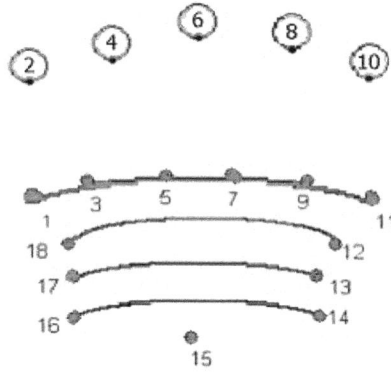

It is a _____.

Show how ♗ can reach the square marked by a ✖ in the shortest way by an arrow.

Connecting the dots according to the number sequence starting from 1.

It is a _____.

Fill in each box with a number.

11		13		15		17		19	

21		23		25		27		29	

31		33		35		37		39	

41		43		45		47		49	

51		53		55		57		59	

10		30		50		70		90	

Connecting the dots according to the number sequence starting from 1.

It is a _____.

Show how Rb2 can reach the square marked by a ✕ in the shortest way by an arrow.

Student's Name _____   Date _____

Connecting the dots according to the number sequence starting from 1.

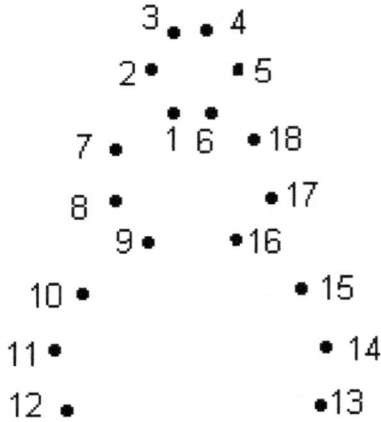

It is a _____.

Fill in each empty box with a number.

↓			↔					
11			15					

21			25					

31			35					

41			45					

10		30		50		70		90	

Student's Name _____   Date _____

Connecting the dots according to the number sequence starting from 1.

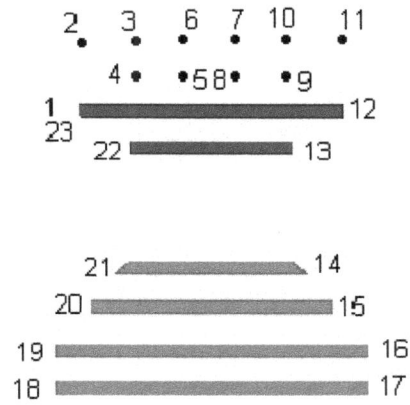

It is a _____.

Show how Rb2 can reach the square marked by a ✕ in the shortest way by an arrow.

Connecting the dots according to the number sequence starting from 1.

It is a _____.

Connect the column 1 to match column 2.

1	♛ ✖		7	A
2	♖ ♗		10	B
3	♙ ✖		9	C
4	♘ ♔		5	D
5	♛ ♝ ♝		4	E
6	✳ ♕ ♙ ♙		3	F
7	♛ ♘ ♙		6	G
8	♗ ♙ ♖		13	H
9	♞ ♙		8	I
10	♕ ♖ ♙		12	J

Student's Name _____   Date _____

Chessboard and math

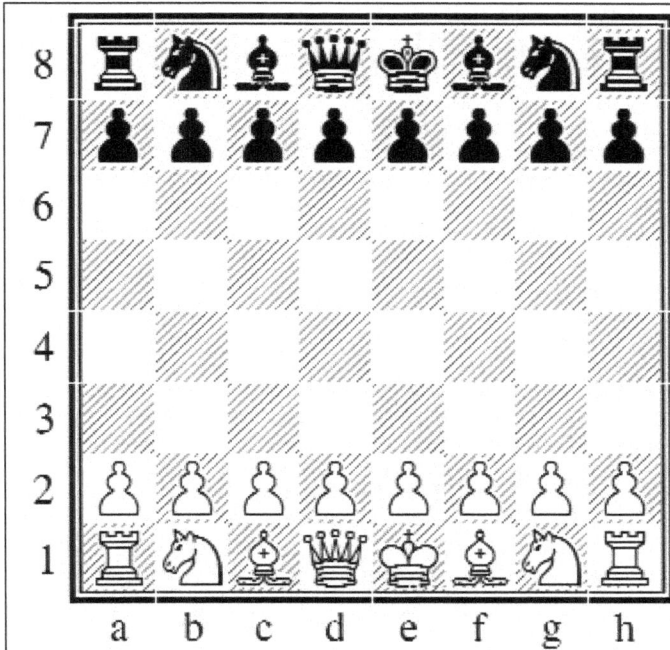

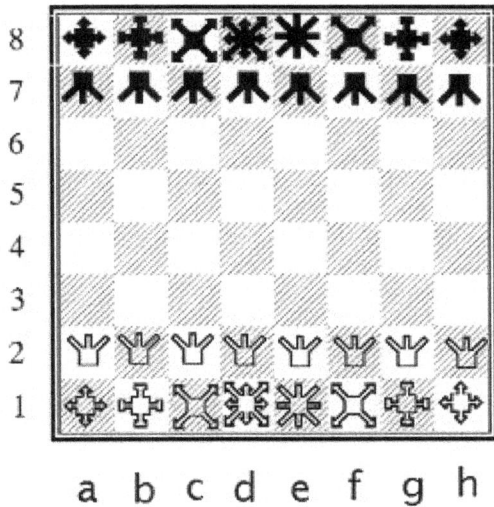

On the left chessboard (without considering the colours of chess squares), there are

<u>8</u> ♙'s and ___ ♟'s .

___ ♖'s and ___ ♜'s.

___ ♘'s and ___ ♞'s. ___ ♗'s and ___ ♝'s.

___ ♕'s and ___ ♛'s.

___ ♔'s and ___ ♚'s.

On the left chessboard (without considering the colours of chess squares), there are

____ ✚'s,

____ ♕'s,

____ ✳'s,

____ ✜'s,

____ ✕'s.

Chessboard notation and coordinates

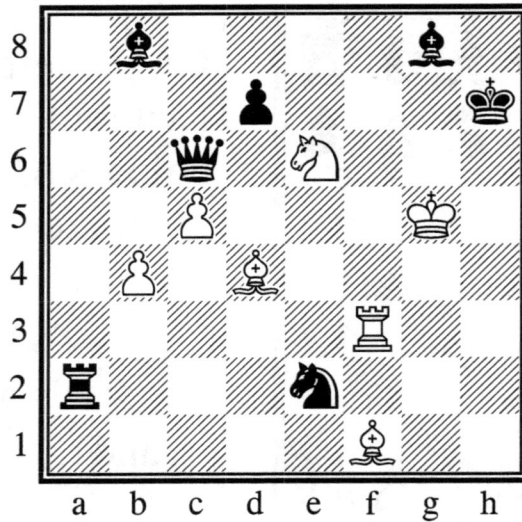

Use the above chess set-up to fill in each blank with a number.

There are _____ ♖ , _____ ♜ , _____ ♗ , _____ ♘ , _____ ♙

_____ ♞ _____ ▨ , _____ ♛ , and _____ ♚ .

Use the above chess set-up to fill in each blank with chess notation.

♛ is on c _____	♗ is on f _____
♚ is on g _____	♝ is on b _____
♗ is on _____	♘ is on _____
♞ is on _____	♜ is on _____
b4 is a _____	d7 is a _____

Student's Name _____ Date _____

Chessboard notation and chess moves

moves from c3 to c7 then to h7, to h5, c5, then back c3. Include c3, how many black squares has

passed by _____. How many white squares has it passed by? _____

Count the number of squares on each rank passed from the left rook to the right rook.

Rank	# of squares moved
8	4
7	?
6	?
5	?
4	?
3	?
2	?
1	?

Chess moves and spatial relation

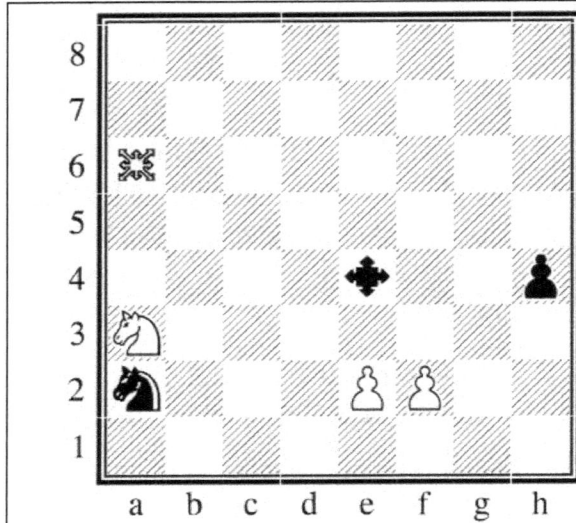

Move ♖ according to the instructions below.

Re4 ⇕ 2 is _____.

Re4 ⇔ 1 is _____.

Re4 ⇔ 4 is _____.

Re4 ⇔ 3 is _____.

Move ♘ according to the instructions below.

Na3 ⊥ is _____,

Na3 ⊥ is _____,

Na3 ⊥ is _____,

Na3 ⊥ is _____.

© 2007 – 2020   Frank Ho, Amanda Ho   All rights reserved. www.homathchess.com
Student's Name _____   Date

Chess movement and spatial relation

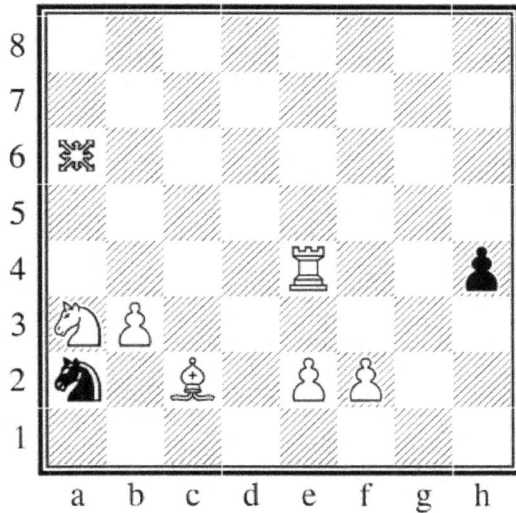

Move ♗c2, according to the instructions below.

⤨ 1 is _____,

The ⤨2 is _____,

then ⤨3 is _____,

then ⤨3 is _____,

then finally ⤨2 is _____.

Describe the shortest path for Bc2 to reach f1 by highlighting the direction on each of the following ⤨ .

⤨ _____ (write moves in chess notation.)

⤨ _____

⤨ _____

⤨ _____

⤨ _____

Chess moves and spatial relation

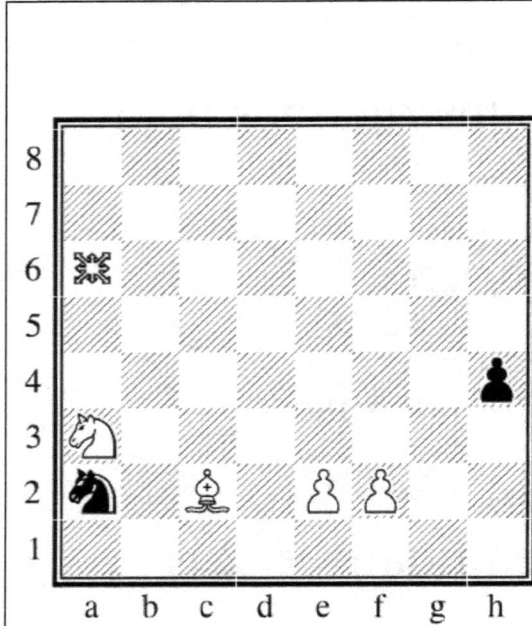

Move ♗ according to the instructions below.

Bc2 ⤢ 2 is _____.

Bc2 ⤢ 1 is _____.,

Bc2 ⤢ 4 is _____

Bc2 ⤢ 1 is _____.

Bc2 ⤢ 1 is _____.

Student's Name _____

Date _____

Chess moves and spatial relation

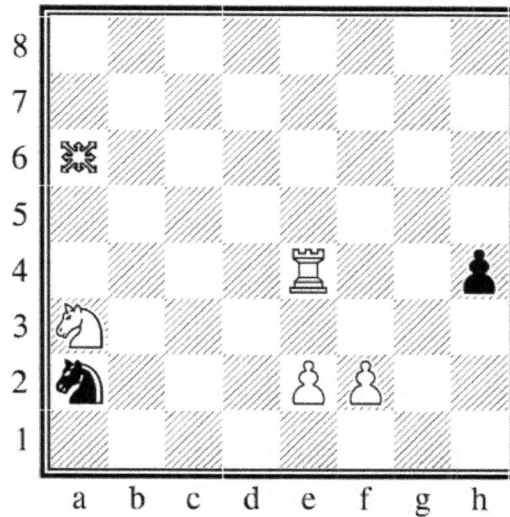

Move chess pieces according to the instructions below.

Re4 ⤢ to g4 will require ♖ moving ___ squares.

Re4 ⤢ to b4 will require ♜ moving ___ squares.

Re4 ⤢ to e7 will require ♜ moving ___ squares. Qa6 ✳ to h6 will

require ♕ moving ___ squares.

Mark the safe squares ✳ from c1 to c8 with a √.

Mark the safe squares ✳ from c8 to h3 with a √.

Student's Name _____   Date _____

Chessboard diagonals, ranks, files

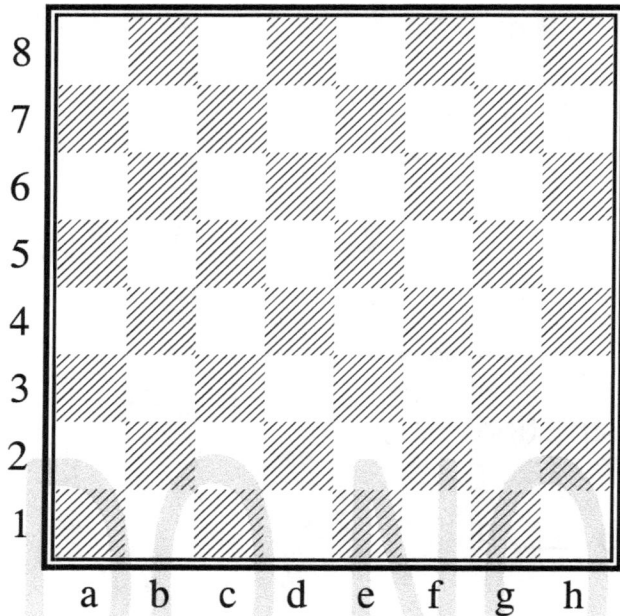

8	
7	
6	
5	
4	
3	
2	
1	
	a b c d e f g h

- Draw the 2 main diagonals of the left chess diagram.

- One of them is from a1 to _____ and the other main diagonal is from _____ to _____.

- Do these two main diagonals meet on the central point? _____.

- On a chessboard, there are 8 ranks. These ranks are 1, 2, 3, ____, ____, ____, ____, and ____.

- There are 8 files. These files are a, b, c, ____, ____, ____, ____, and ____.

Student's Name _____　　Date _____

Chessboard

Name the squares in algebraic notation where the can safely move to

_____.

Name the squares in algebraic notation where the can safely move to

_____.

Name the squares in algebraic notation where the can safely move to

_____.

Name the squares in algebraic notation where both the and can safely move to

_____.

Student's Name _____ Date _____

Chessboard

Chessboard

The number of squares where Rb2 can safely move to is _____.

Chessboard

The number of squares where Nf4 can safely move to is _____.

Student's Name _____   Date _____

Chessboard

Name the squares in algebraic notation where both the ⚓ and ⤧ can safely move to

_____.

Name the squares in algebraic notation where both the ✹ and ✕ can safely move to

_____.

Name the squares in algebraic notation where all ✚ and ✹ and ✕ can safely move to

_____.

A spatial relation on information retrieving, classifying, interacting, sorting, counting, ordering

Spatial relation and sorting

Chessboard	
	The number of ■'s controlled by ♛ is _____ The number of •'s controlled by ♛ is _____ The number of ▲'s controlled by ♛ is _____
Chessboard	
	The number of ■'s controlled by ♛ is _____ The number of •'s controlled by ♛ is _____ The number of ▲'s controlled by ♛ is _____

Student's Name _____   Date _____

	Direction	Direction	Direction	Direction
♛	Colour red	Colour green	Colour blue	Colour black
♞	Colour red	Colour green	Colour blue	Colour black
♞	Colour red	Colour green	Colour blue	Colour black

Student's Name _____　　　　　　Date _____

Spatial relation and directions

	Direction	Direction	Direction	Direction
3 4 5 7 9 / 1 _ _ _ 6 / 4 _ ♛ _ 4 / 6 _ _ _ 5 / 7 6 8 4 3	List the numbered square on Queen's next move as indicated ☐	List the numbered square on Queen's next move as indicated ☐	List the numbered square on Queen's next move as indicated ☐	List the numbered square on Queen's next move as indicated ☐.
3 4 5 7 9 / 1 _ _ _ 6 / 5 _ ♞ _ 6 / 9 _ _ _ 8 / 7 6 8 4 3	List the numbered square on knight's next move as indicated ☐.	List the numbered square on knight's next move as indicated ☐.	List the numbered square on knight's next move as indicated ☐.	List the numbered square on knight's next move as indicated ☐.
3 4 5 7 9 / 1 _ _ _ 6 / 5 _ ♞ _ 5 / 3 _ _ _ 4 / 7 6 8 4 3	List the numbered square on knight's next move as indicated ☐.	List the numbered square on knight's next move as indicated ☐.	List the numbered square on knight's next move as indicated ☐.	List the numbered square on knight's next move as indicated ☐.

何数棋谜　低年级棋谜式数学

Student's Name _____　　　　　Date _____

Spatial relation and information processing

	Calculating follows directions
	Calculate the result of the numbered squares as directed above _____.
	Calculate the result of the numbered squares as directed above _____.
	Calculate the result of the numbered squares as directed above _____.

Student's Name _____   Date _____

Spatial relation and connecting

	Connect 4
3 4 5 7 9 1 _ _ _ 6 4 _ ♛ _ 4 6 _ _ _ 5 7 6 8 4 3	 Connect 4 numbers using the above 4 directions in the order of left to the right.
3 4 5 7 9 1 _ _ _ 6 5 _ ♛ _ 6 9 _ _ _ 8 7 6 8 4 3	 Connect 4 numbers using the above 4 directions in the order of left to the right.
3 4 5 7 9 1 _ _ _ 6 5 _ ♞ _ 5 3 _ _ _ 4 7 6 8 4 3	 Connect 4 numbers using the above 4 directions in the order of left to the right

46

Student's Name _____ Date _____

Spatial relation, classifying, sorting, counting

8	✛	✸	♛	✛	✸	✛	♛	✛
7	♛	♛	✛	♛	✛	✖	✛	♛
6	✖	✸	✛	✸	✸	✛	♛	✛
5	♛	✖	♛	♛	✛	✖	✛	✖
4	✖	✖	✸	✸	✛	♛	✸	♛
3	✛	✸	✖	✖	✸	✖	✛	♛
2	✛	✖	♛	✸	✛	♛	✸	✛
1	✖	♛	♛	✖	✸	✖	✛	♛
	a	**b**	**c**	**d**	**e**	**f**	**g**	**h**

Find chess pieces within the shaded squares.	♛	✖	✸	✛
(top-left shaded)				
(top-right shaded)				
(bottom-left shaded)				
(middle-bottom shaded)				

Spatial relation, classifying, sorting, counting

Chess moves at the following locations.	How many of the following chess pieces are there on the path of chess moves?			
	⛊	✗	✸	✦
Re4				
Bf3				
Qg2				
Ra3				
f4				
Qd4				

Spatial relation and sorting

1	1	10	1	5
10	5	1	5	1
5	✖	10	10	1
10	1	3	10	1
5	5	1	10	5

The number of 10's controlled by ♛ is

_____

The number of 5's controlled by ♛ is

_____

The number of 1's controlled by ♛ is

_____

Chessboard

The number of ■'s controlled by ♛ is

_____

The number of ●'s controlled by ♛ is

The number of ▲'s controlled by ♛ is

_____

Chessboard

The number of ■'s controlled by ♛ is

_____

The number of ●'s controlled by ♛ is

_____

The number of ▲'s controlled by ♛ is

_____

Student's Name _____   Date _____

Spatial relation and interaction

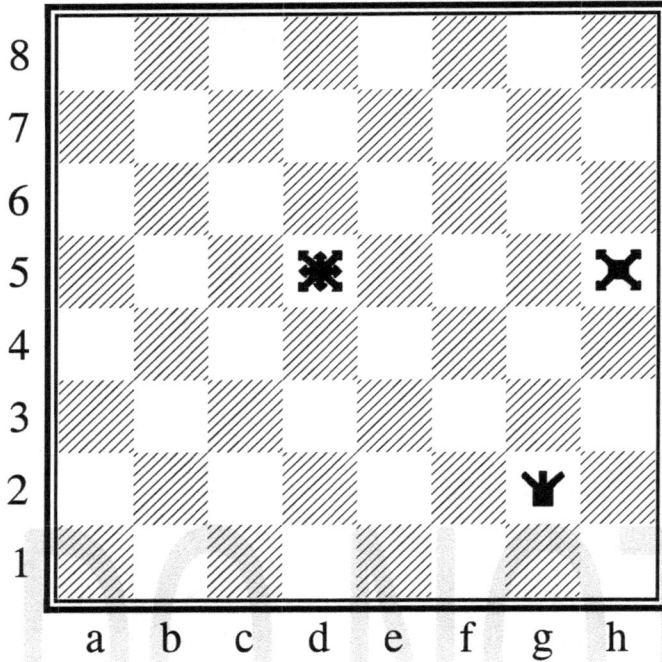

Mark squares which can be controlled by

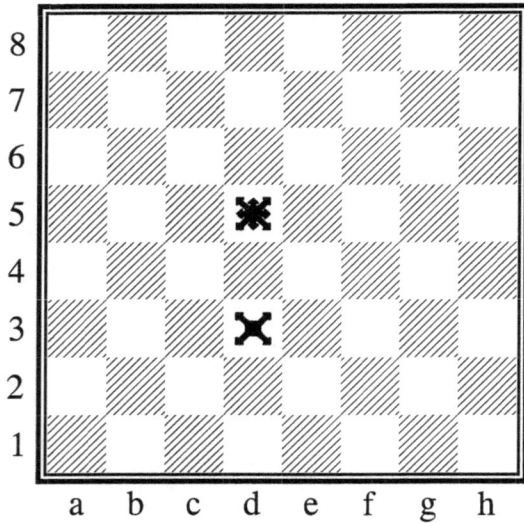

♕ , ✚ , ✗ , and ✸ .

Mark squares, which can be controlled by both ✗ and ✸ .

Spatial relation and ordering

(a). List all occupied squares where ✳ can safely reach. List these squares in points.

_____

(b) Arrange them in order from the smallest to the largest.

_____.

(a). List all occupied squares where ✳ can safely reach. List these squares in points.

_____

(b) Arrange them in order from the smallest to the largest.

_____.

Spatial relation and sorting

	Find Answers
<table><tr><td>6</td><td>4</td><td>6</td><td>3</td><td>3</td></tr><tr><td>3</td><td>8</td><td>8</td><td>3</td><td>0</td></tr><tr><td>3</td><td>4</td><td>♘</td><td>2</td><td>5</td></tr><tr><td>3</td><td>4</td><td>6</td><td>3</td><td>2</td></tr><tr><td>6</td><td>7</td><td>1</td><td>7</td><td>2</td></tr></table>	Circle all the numbers that can be reached in the knight's next move and list them from the smallest to the largest.
<table><tr><td>6</td><td>4</td><td>6</td><td>3</td><td>6</td></tr><tr><td>4</td><td>1</td><td>7</td><td>3</td><td>5</td></tr><tr><td>4</td><td>9</td><td>♞</td><td>6</td><td>7</td></tr><tr><td>3</td><td>8</td><td>8</td><td>9</td><td>6</td></tr><tr><td>3</td><td>2</td><td>5</td><td>1</td><td>4</td></tr></table>	Circle all the numbers that can be reached in the knight's next move and list them from the smallest to the largest. _____ _____

何数棋谜　低年级棋谜式数学

Student's Name _____   Date _____

Spatial relation

	↑ →	← ↑
	Use the above direction to find out where can ❈ safely go? Write the names of squares in chess notation. _____	Use the above direction to find out where can ❈ safely go? Write the names of squares in chess notation. _____
	Use the above direction to find out where can ❈ safely go? Write the names of squares in chess notation. _____ f1, f3, f5, g2	Use the above direction to find out where can ❈ safely go? Write the names of squares in chess notation. _____ f1, f3, f5, e3

Student's Name _____   Date _____

Dividing a mini-chessboard

How many ways can you divide the following square into 4 parts such that each ⊥ occupies in each part?

何数棋谜　低年级棋谜式数学

Chess pieces moving ✦ directions

If the following ♖ can make one move from e4 freely in all possible directions up to a maximum of 4 directions, then what possible directions are there? Draw each direction.

		Move 1 direction. 4 ways One of the answers is shown below.	Move 2 directions. 6 ways One of them is shown below.	Move 3 direction 4 ways One of them is shown below.	Move 4 directions. 1 way

Chess pieces moving ✕ directions

If the following ♝ can move from d5 freely in all possible directions up to 4 directions, then what possible directions are there?

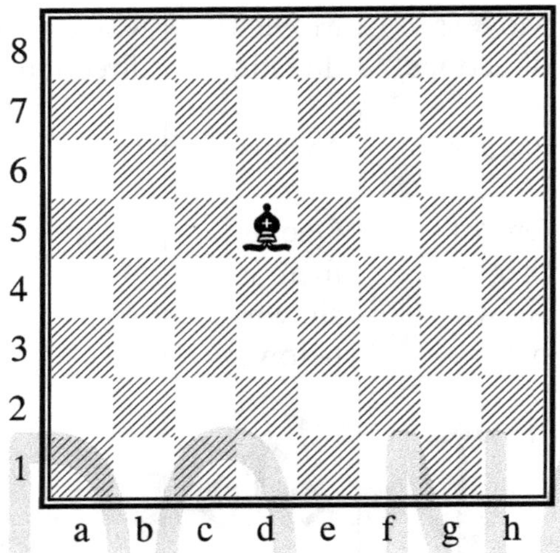

	Move 1 direction.	Move 2 directions.	Move 3 directions.	Move 4 directions.
		5 ways	4 ways	
		One of them is shown below.	One of them is shown below.	

Student's Name _____ Date _____

Counting and spatial relation

Count the number of chess pieces totally within in each shaded box and fill in a number in each ___.

Counting and spatial relation

Count the number of chess pieces totally within each shaded box and fill in a number in each ____.

Student's Name _____   Date _____

Rook path

	Starting at a corner, how many minimum moves must rook take to pass every square?
	Starting at a corner, how many minimum moves must rook take to pass every square?
	Starting at a corner, how many minimum moves must rook take to pass every square?
	Starting at a corner, how many minimum moves must rook take to pass every square?

Rook path

How many ways can the ♖ travel to the ✜ bypassing each square only once? Mark the path by lines.

何数棋谜　低年级棋谜式数学

Magic Number in rook's move

Fill in numbers of 1, 2, 3, 4, 5 to the right boxes in **rook**'s moves such that all 3 numbers added up to be the same.	
Fill in numbers of 2, 4, 6, 8, 10 to the right boxes in **rook**'s moves such that all 3 numbers added up to be the same.	
Fill in numbers of 1, 3, 5, 7, 9 to the right boxes in **rook**'s moves such that all 3 numbers added up to be the same.	

Student's Name _____　　　　　　　　Date _____

Magic Number in rook's or bishop's move

Fill in numbers of 11, 12, 13, 14, 15 to the right boxes in **bishop**'s moves such that all 3 numbers added up to be the same.

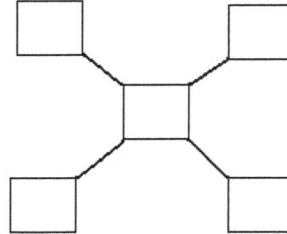

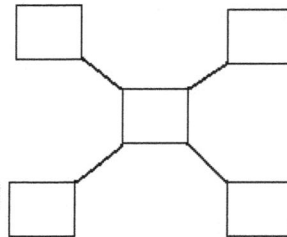

Fill in numbers of 3, 5, 7, 9, 11 to the right boxes in **bishop**'s moves such that all 3 numbers added up to be the same.

Fill in numbers of 3, 5, 7, 9, 11 to the right boxes in **bishop**'s moves such that all 3 numbers added up to be the same.

© 2007 − 2020   Frank Ho, Amanda Ho

Student's Name _____   Date _____

Chess and geometry

Connecting all points to form lines

Connect 2 points.	Connect 2 points.	Connect 2 points.	Connect 2 points.
Connect 3 points.	Connect 3 points.	Connect 3 points.	Connect 3 points.
Connect 3 points.	Connect 3 points.	Connect 3 points.	Connect 3 points.
Connect 3 points.	Connect 3 points.	Connect 3 points.	Connect 3 points.

Student's Name _____   Date _____

Chess Pattern and Geometry

Chess Diagram	Question	Geometry	Geometry
	Can the black queen move? _____	Show by an arrow to move one point, so 3 points are on a horizontal or vertical line.	Show by an arrow to move one point, so 3 points are on a horizontal or vertical line.
	Can the black queen move? _____	Connect 3 points such that all 3 points are on a line.	Connect 3 points such that all 3 points are on a line.
	Can the black queen move? _____	Connect 3 points such that all 3 points are on a line.	Connect 3 points such that all 3 points are on a line.

Counting

♕			
(9)			
♙	✕	✳	✜
0	0	1	0

♖			
♙	✜	✳	✜
0	0	0	1
2	1	0	0
5	0	0	0

Student's Name _____   Date _____

Counting

Chess diagram	
	The number of squares where Qf2 can safely move to is _____.
Chess diagram	
	The number of squares where Bd3 safely move to is _____.

Student's Name _____ Date _____

Counting

Chess diagram	
	The number of squares where Qf2 can safely move to is _____.
Chess diagram	
	The number of squares where ♛ safely move to is _____.

Student's Name _____ Date _____

Counting

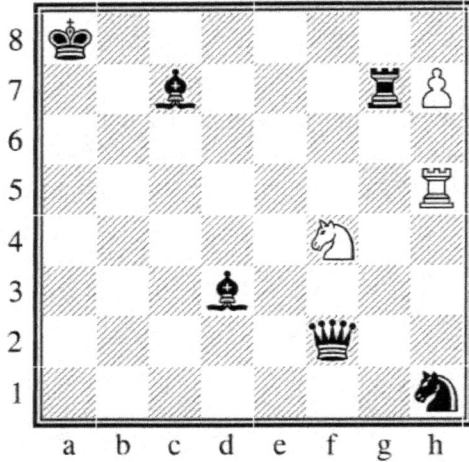

Chess diagram	
	The total points shown on the number of squares where Qf2 can safely move to is _____.
Chess diagram	
	The total points shown on many squares where Bd3 safely move to is _____.

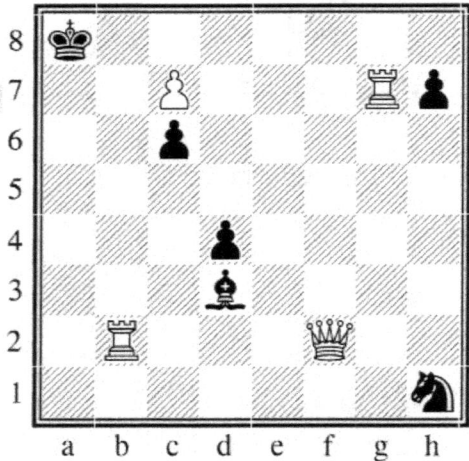

Student's Name _____   Date _____

Venn diagram and chess set

Venn Diagrams

Find the common elements between A and B.

A= {1, 2, 3, 4, 5}
B= {1, 3, 11, 12, 14}

A ∩ B = _____

Chessboard

Find the squares that are controlled by both rook and queen.

♕ = {1, 2, 3, 4}
♖ = {1, 2, 3, 4}

♖ ∩ ♕ = {____, ____} How is Venn diagram different from chess set? _____

Student's Name _____   Date _____

Trace rook's and queen's common squares and then find their answer.

2	2	♛
2	3	2
♜	1	1
1	0	2

What is the total value of squares where both ♛ and ♜ can control?

⇕ + ⊠ + ⇕

Answer: _____

何数棋谜　低年级棋谜式数学

Pattern

1. Replace each ? with a chess piece.

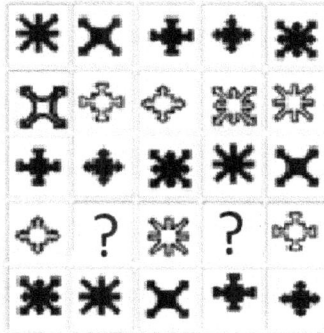

2.

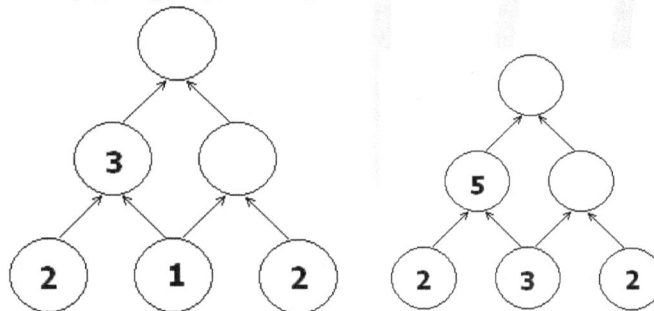

3. At the end of a chess game, White had 4 points, and Black has 3 more points than White. How many points does Black have?

1. Find the pattern.

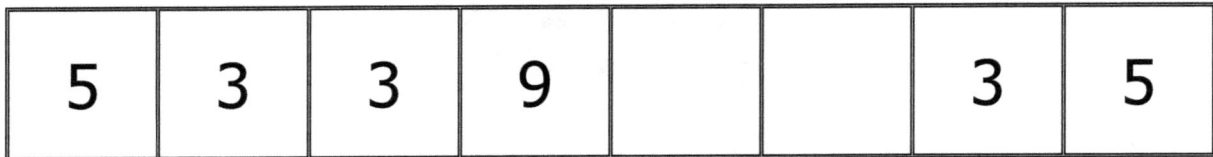

♜	♞	♝	♛	♚			♜

5	3	3	9			3	5

K	T	P		G	P		

2. Finish the following diagram.

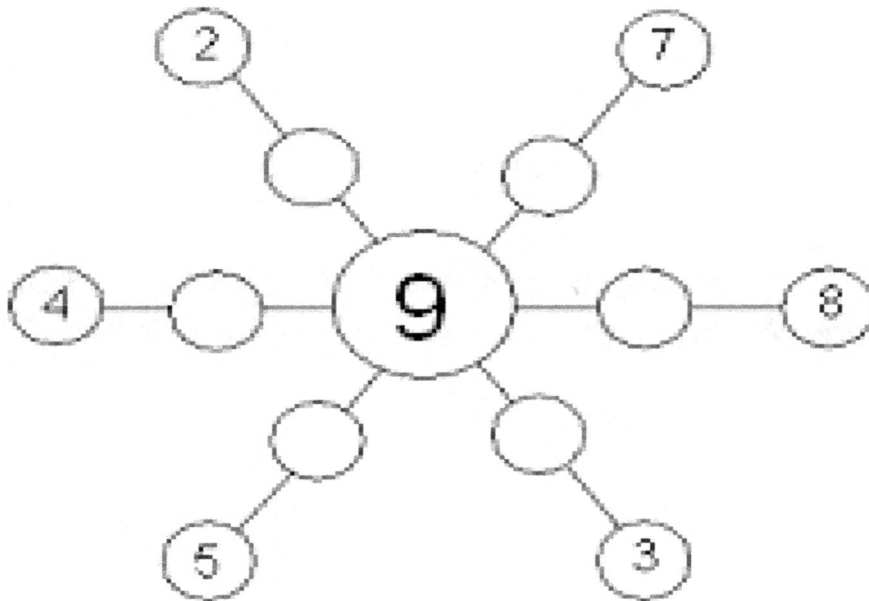

1. Finding pattern and filling in blanks

   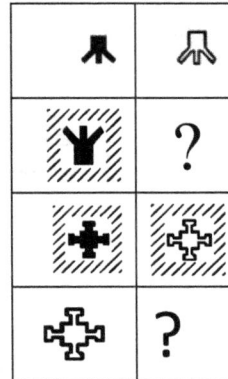

2. Trace rook's moves and find the answer. In this problem, ♖ moves one square always.

2	2	2
2	3	2
♖	1	1
1	0	2

= _____ = _____

= _____

_____ = _____

Finding a pattern

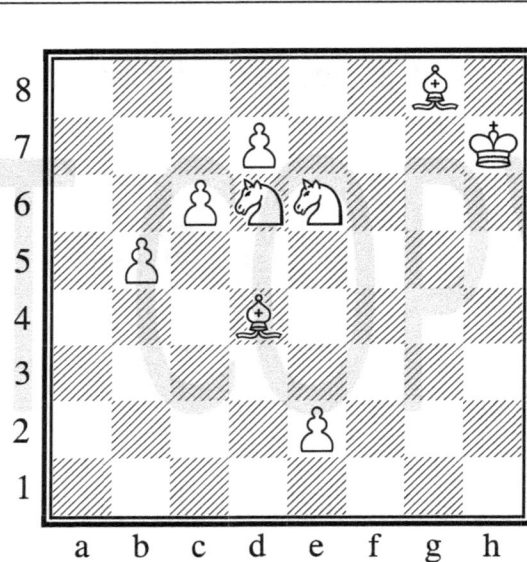

The total value of the left diagram + the values of the right diagram is _____

Student's Name _____   Date _____

1. Colour every 3rd of the following ♛'s. How many ♛'s are uncoloured?

♛ ♛ ♛ ♛ ♛ ♛ ♛ ♛ ♛ ♛

2. Colour every 5th of the following ♞'s. How many ♞'s are uncoloured?

♞ ♞ ♞ ♞ ♞ ♞ ♞ ♞ ♞ ♞

3. What number is 2 more than the number of the following ♙ 's?

Answer: _____

♙ ♙ ♙ ♙ ♙ ♙ ♙ ♙ ♙ ♙ ♙ ♙ ♙ ♙ ♙ ♙ ♙

4. Fill in each blank with a number.

♛	✚	✢		❋		13		17	
2	4	6		10		14		18	

21	22	23		25				

10	20			50		70		

11	21	31			61				101

1. Finding the pattern and filling in each ? by a number

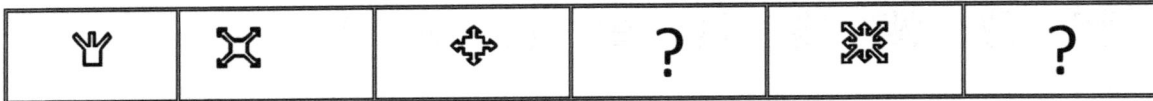

♛	⌗	✛	?	❊	?

2. Fill in each blank with a number. The same shape means the same number for each problem.

1. **0 +** ☐ **+** ☐ **= 10**

2. **2 +** ☐ **+** ☐ **= 10**

3. **4 +** ☐ **+** ☐ **= 10,**

4. **6 +** ☐ **+** ☐ **= 10**

5. **8 +** ☐ **+** ☐ **= 10**

3. Fill in numbers of 1, 2, 3, 4, and 5 to the right boxes in rook's moves such that all 3 numbers added up to be the same.

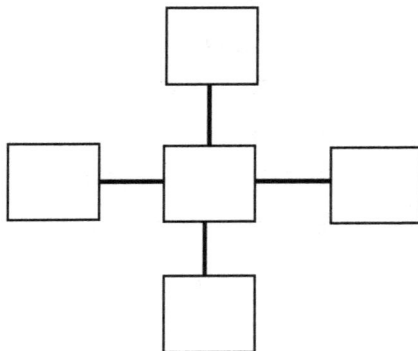

1. Can you find letters to replace the question marks?

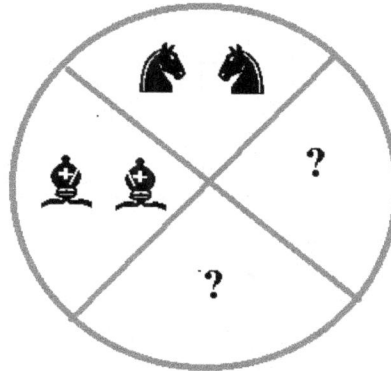

2. Find all the combinations that can make the same value as a rook. One is done for you.

♙	✚	✳	✦
0	0	0	1
2	3		0
5	0		0

3. At the end of a chess game, White had a ♛, a ♜, a ♞ and a ♚ left. How many points did White have?

Student's Name _____   Date _____

1. Trace rook's and queen's common squares and then find their answer.

5	6	♛	9
2	8	2	1
♜	1	7	2
1	0	5	3

Answer : ____ + ____ = ____

2. Complete the following table.

6
26
36
56

60	61						

76
86

Draw the next sequence of the diagram.

1. At the end of a chess game, white has a queen, a rook, and a king left. How many points does white have now?

2. At the end of a chess game, white has two rooks and a king left. How many points does white have now?

3. At the end of a chess game, Black has one rook, one bishop, and a king left. White has one more rook than B13lack. How many points does white have now?

Fill in each □ with an answer.

Student's Name _____   Date _____

Complete the next term of each pattern.

10, 100, 1000, _____, _____, _____.

121, 132, 1__, 15__, __ __ __.

231, 242, 2__ __, 26__, __ __ __.

abc, acb, bac, _____, _____, _____.

Complete the following diagrams based on diagram 1.

Complete the next term of each pattern.

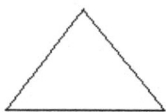

$\frac{1}{5}$	$\frac{2}{5}$	$\frac{3}{5}$	**?** $\frac{4}{5}$

Complete the next term of each pattern.

$\frac{1}{8}$	$\frac{2}{8}$	$\frac{3}{8}$	

Complete the next term of each pattern.

_____.

Complete the next term of each pattern.

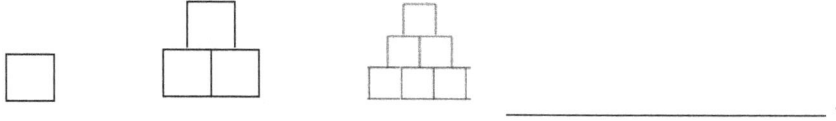

Complete the next term of each pattern.

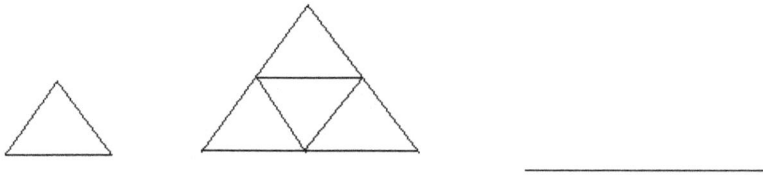

Complete the next term of each pattern.

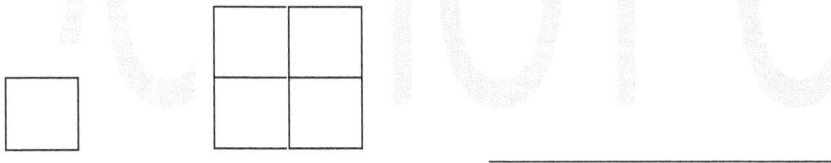

Complete the 4th term as indicated by an arrow.

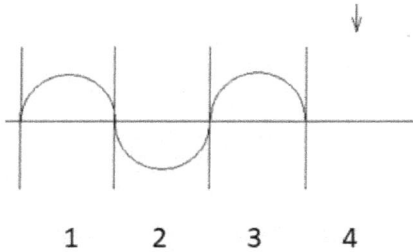

1　　2　　3　　4

Complete the 5th and 6th terms as indicated by an arrow.

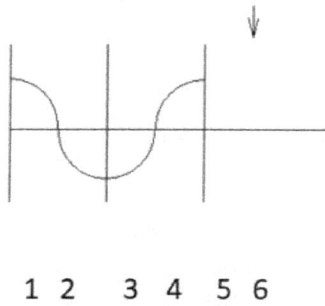

1 2　3　4　5　6

Connect all diagonals.

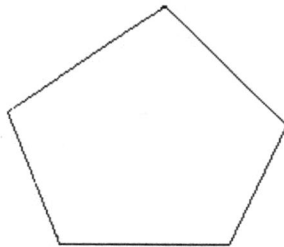

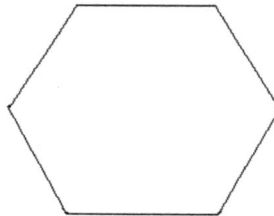

© 2007 − 2020　　Frank Ho, Amanda Ho　　All rights reserved. www.homathchess.com
Student's Name _____ Date _____

Cut circle and triangle into halves.

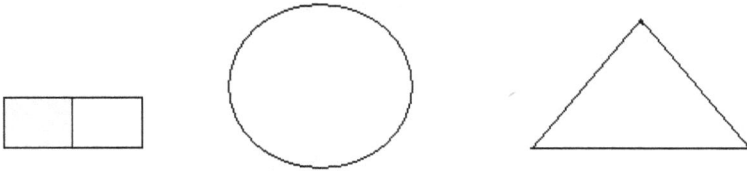

Fill in blanks.

(1,1)　　　　(2,2)　　　　(　,　)　　　　(　,　)
(1,2)　　　　(2,3)　　　　(　,　)　　　　(　,　)

Complete the last diagram.

1. |||||

2. ||||| |||||

3. _____

Student's Name _____   Date _____

Clock face reading

Draw lines to match clock to time and then to chess pieces.

Clock	Time	Chess pieces
	6 O'clock	
	3 O'clock	
	2 O'clock	

何数棋谜 低年级棋谜式数学

Student's Name _____ Date

Clock face reading

Draw lines to match clock to time and then to chess pieces.

Clock	Time	Chess pieces
	5:00	
	7:00	
	9:00	

Number patterns

Find the next 4 numbers.

88, 87, 85, 82, 78, _____, _____, _____, _____ Pattern rule is
_____

1, 1, 2, 3, _____, _____
Pattern rule is _____ A

41, 42, 44, 47, _____, _____ Pattern rule is
_____ A

Number patterns in the table

1	5
2	10
3	15
4	?
5	?

Replace the following each ? by a number

2	5	2
4	♖♖	7
?	♖♖♖	?
8	?	17
?	?	?

Student's Name _____   Date

Graph pattern

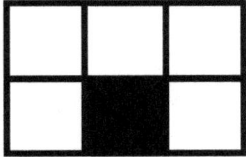

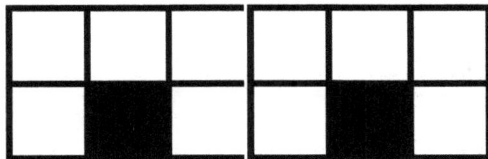

What is the next graph?

Draw the next shape in the following table and complete a T-chart.

Design	# of ♟'s
	5
	8

Grid pattern with 1 attribute changing

Replace each ? with a chess piece.

Grid pattern with 2 attributes changing

Replace each ? with a chess piece.

Grid pattern with 3 attributes changing

Replace each ? with a chess piece.

Student's Name _____ Date _____

Pattern with 3 attributes changing

Replace each ? with a chess piece.

Pattern with transformations

What is the next graph?

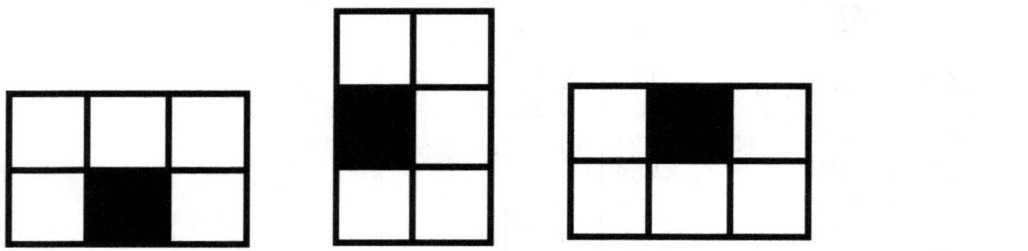

Student's Name _____   Date _____

Computation

❊ ♔ = 0, ♕ ♙ = 1, ✖ ♗ = 3, ✚ ♘ = 3, ✚ ♖ = 5, ❊ ♛ = 9

Connecting numbers in order starting from 1 to 124.

		1	2	3	4	15	16	17	18
		8	7	6	5	14	21	20	19
		9	10	11	12	13	22	23	24
		34	33	32	31	30	27	26	25
63	62	61	60	35	36	37	38	29	28	43	44
64	65	66	59	58	53	52	39	40	41	42	45
73	72	67	68	57	54	51	50	49	48	47	46
74	71	70	69	56	55	90	91	92	93	94	95
75	80	81	86	87	88	89	100	99	98	97	96
76	79	82	85	104	103	102	101				
77	78	83	84	105	118	119	120				
110	109	108	107	106	117	122	121				
111	112	113	114	115	116	123	124				

93

Student's Name _____   Date _____

Number sequence

♔ = 0, ♙ = 1, ♗ = 3, ♘ = 3, ♖ = 5, ♕ = 9

Fill in each empty box with a number.

| ♕ | 2 | | ♖ | 6 | 8 | | 11 | |

| ♕ | | ♗ | | ♖ | | 7 | | ✹ | |

| | 2 | | 4 | | 6 | | 8 | | 10 | |

| 11 | 21 | | | 51 | 61 | | 81 | | | 111 |

| 11 | | 13 | | 15 | | 17 | | 19 | |

| | 20 | | 40 | | 60 | | 80 | | 100 |

Student's Name _____   Date _____

Number sequence

✻ ♔ = 0, ♕ ♙ = 1, ✖ ♗ = 3, ♣ ♘ = 3, ♦ ♖ = 5, ✻ ♕ = 9

Fill in each ☐ with a number.

1	♦	6	✻	♕
2	4	7	4	2
✖	♣	☐	☐	☐
4	2	☐	☐	☐
♦	♕	☐	☐	☐

Student's Name _____　　　　Date _____

Connecting two sides with the same number of total points

1	✚ ✚ ✚ ✚ ✚		✳ ✳ ✳ ✳ ✳	A
2	✧ ✧ ✧		✖ ✖	B
3	✳ ✳		✚ ✚ ✚ ✚ ✚	C
4	✖ ✖ ✖ ✖ ✖		✦ ✦ ✦	D
5	♟ ♟ ♟		♟ ♟ ♟	E

Student's Name _____   Date _____

Connect numbered dots up to 100.

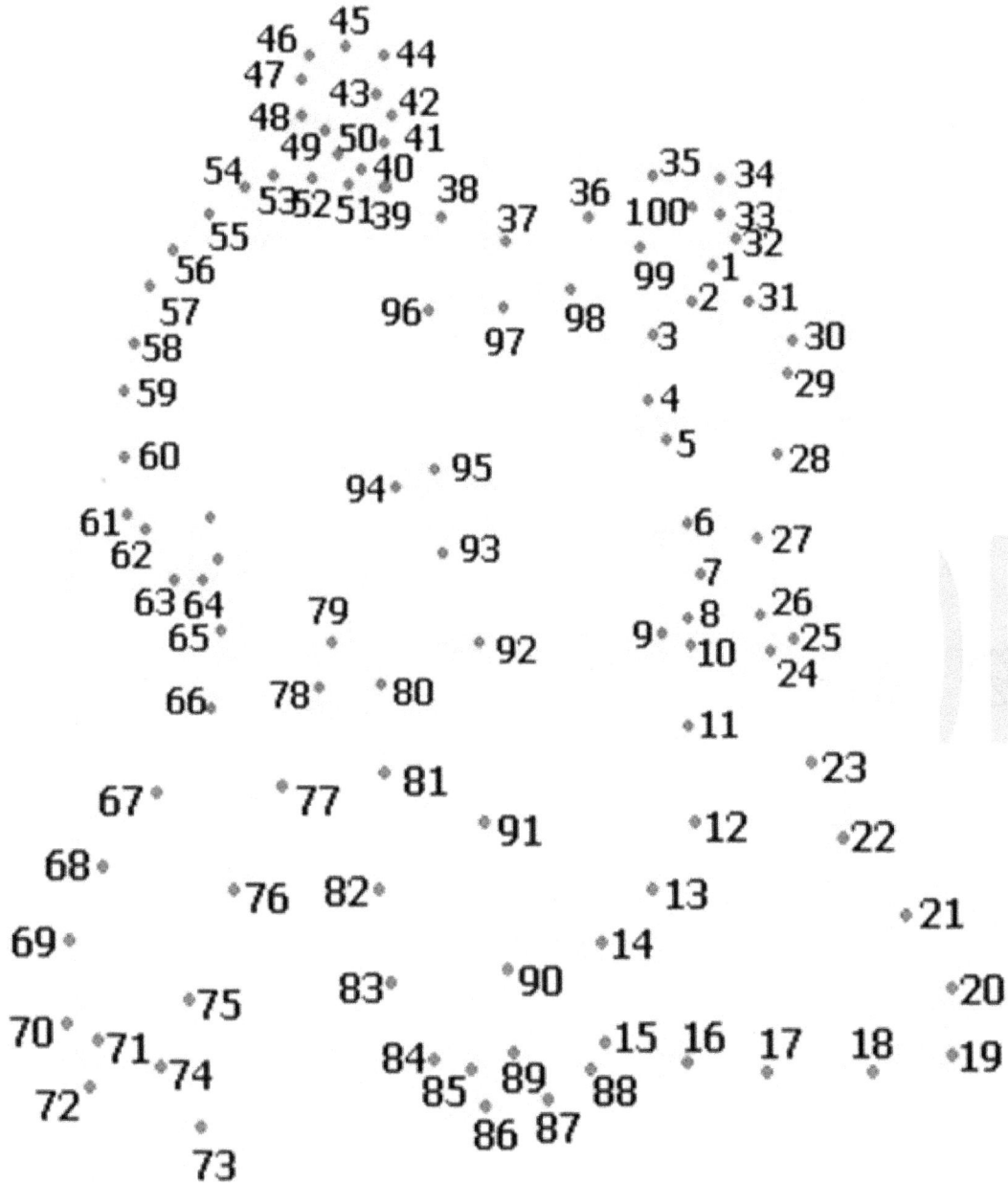

Tallying

1. Tally the following chess pieces.

(‖‖‖ /)	( )
( )	( )

2. Count the number of each chess piece.

Student's Name _____　Date _____

Count the number of chess pieces and tally them.

1. Tallying the following data

♟ or ♛	♕ or ✳
(rook path ~~////~~ /)	(_____)
♞ or ✠	✗ or ♝
(_____)	(_____)

2. Pattern

23	24	25		
	34			
		45		
			57	

Student's Name _____   Date _____

Matching

1. Connect two sides by the same number of chess pieces (not points).

2. Fill in the following blanks.

1	2		5	6		8			11	

1		3		5		7		9	

	2		4		6		8		10

11	21			51	61		81		

11		13		15		17		19	

	20		40		60		80		100

101

Student's Name _____   Date _____

Ordering

1. Fill in the missing number.

1	2	3					9	
	12		14			18		
			25				29	30
31				37				
	43	44	45					
					58	59	60	
	62	63		67				
			75	76				
	83	84				88		
91			96	97				

2. From the table above, find the missing number.

13	14	15		
64			67	

31	32		
		44	

Student's Name _____   Date _____

Chessboard mathematics by using its direction, geometry translation, chess notation

5, 10, 15, _____, _____, _____, _____, _____, _____, _____, _____, _____, _____,

_____, _____, _____, _____, _____, _____,

2, 4, 6, ____, _____, _____, _____, _____, _____

4, 8, 12, _____, _____, _____

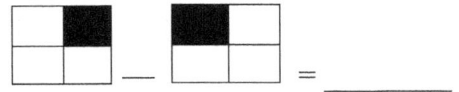 = _____

3, 6, 9, _____, _____, _____, _____, _____, _____, _____, _____, _____,

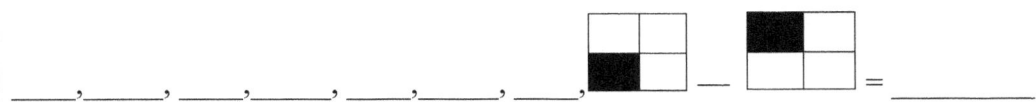 = _____

Direction, geometry translation, chess notation

5, 10, 15, _____, _____, _____, _____, _____, _____, _____, _____, _____,

_____, _____, _____, _____, _____, _____, _____, _____,

2, 4, 6, ____, ____, ____, ____, ____

[grid] + [grid] = _____

3, 6, 9, _____, _____, _____, _____, _____, _____, _____, _____, _____,

_____, _____, _____, _____, _____, _____, _____,

[grid] + [grid] = _____

Student's Name _____   Date _____

Direction, geometry translation, chess notation and computing

10, 20, 30, _____, _____, _____, _____, _____, _____, _____, _____, _____,

_____, _____, _____, _____, _____, _____,

— = _____

3, 6, 9, _____, _____, _____, _____, _____, _____, _____, _____, _____, _____,

_____, _____, _____, _____, _____, _____, _____,

— = _____

Student's Name _____   Date _____

Direction, geometry translation, chess notation

a5	b5	c5	d5	e5
a4	b4	c4	d4	e4
a3	b3	✜	d3	e3
a2	b2	c2	d2	e2
a1	b1	c1	d1	e1

A rook above moves north 2 squares, east 2squares, south 3 squares, west 4 squares and north 3 squares, where is it now?

Answer in algebraic notation _____

Direction, geometry translation, chess notation and computing

5, 10, 15, _____, _____, _____, _____, _____, _____, _____, _____, _____,

_____, _____, _____, _____, _____, _____,

3, 6, 9, _____, _____, _____, _____, _____, _____, _____, _____, _____,

_____, _____, _____, _____, _____, _____,

Student's Name _____   Date _____

Direction, geometry translation, chess notation and computing

5, 10, 15, _____, _____, _____, _____, _____, _____, _____, _____, _____, _____,

_____, _____, _____, _____, _____, _____,

◻◼ / ◻◻ ◻◻ / ◼◻ — _____ = _____

3, 6, 9, _____, _____, _____, _____, _____, _____, _____, _____, _____, _____,

_____, _____, _____, _____, _____, _____, _____,

◻◻ / ◻◼ ◼◻ / ◻◻ — _____ = _____

Student's Name _____　　　　　Date _____

Nested chessboards

Step 1 Select	Step 2 Operate according to the shaded areas
	_____ + _____ + _____ + _____ = ____ ✦ ✦
	_____ + _____ + _____ + _____ = ____ ♛

Nested chessboards

♛	2	♛	2	5	✦	5	✦
♛	♛	♛	♛	✦	5	✦	5
♛	2	♛	2	5	✦	5	✦
♛	♛	♛	♛	✦	5	✦	5
✕	✕	✕	✕	✖	♛	✖	♛
✕	2	✕	✕	✖	♛	✖	♛
✕	✕	✕	✕	✖	♛	✖	♛
✕	2	✕	✕	✖	♛	✖	♛

Step 1 Select	Step 2 Operate according to the shaded areas
+ + −	✦ ✦ ✖ ✖ ♛ ♛ ♛ ♛ ♛ ✕ ✕ ✕ ✕
− =?	

Student's Name _____ Date _____

Adding 1, 2, 3

✳ ♔ = 0, ♕ ♙ = 1, ✪ ♗ = 3, ♧ ♘ = 3, ♣ ♖ = 5, ✳ ♛ = 9

1 + 3 = ___	4 + 1 = ___
5 + 1 = ___	1 + 4 = ___
8 + 1 = ___	7 + 1 = ___

♕	♣	♧	✪	✳	✳
1	5	3	3	0	9

$$\begin{array}{c} ♕ \\ +\ ✳ \\ \hline \square \end{array} \qquad \begin{array}{c} 3 \\ +1 \\ \hline \square \end{array} \qquad \begin{array}{c} 2 \\ +✳ \\ \hline \square \end{array} \qquad \begin{array}{c} 2 \\ +\ ♕ \\ \hline \square \end{array} \qquad \begin{array}{c} 3 \\ +1 \\ \hline \square \end{array}$$

$$\begin{array}{c} ♕ \\ +4 \\ \hline \square \end{array} \qquad \begin{array}{c} 7 \\ +1 \\ \hline \square \end{array} \qquad \begin{array}{c} 6 \\ +♙ \\ \hline \square \end{array} \qquad \begin{array}{c} 7 \\ +✳ \\ \hline \square \end{array} \qquad \begin{array}{c} 5 \\ +♙ \\ \hline \square \end{array}$$

$$\begin{array}{c} 6 \\ +♙ \\ \hline \square \end{array} \qquad \begin{array}{c} ♕ \\ +9 \\ \hline \square\ \square \end{array} \qquad \begin{array}{c} ♙ \\ +3 \\ \hline \square \end{array} \qquad \begin{array}{c} 4 \\ +♙ \\ \hline \square \end{array} \qquad \begin{array}{c} 1 \\ +8 \\ \hline \square \end{array}$$

Student's Name _____   Date _____

Adding 1

♚ = 0, ♛ = 1, ♝ = 3, ♞ = 3, ♜ = 5, ♛ = 9

1. Fill in blanks (adding 1).

♟ + ♟ $1 + 1 = 2$	♟ ♟ ♟ + ♟ $3 + 1 = \underline{\quad}$
♟ ♟ + ♟ ♟ ♟ $\underline{\quad} + \underline{\quad} = \underline{\quad}$	♟ ♟ ♟ + ♟ ♟ ♟ ♟ $\underline{\quad} + \underline{\quad} = \underline{\quad}$
♟ ♟ ♟ ♟ + ♟ ♟ ♟ ♟ ♟ $\underline{\quad} + \underline{\quad} = \underline{\quad}$	♟ ♟ ♟ + ♟ ♟ ♟ ♟ $\underline{\quad} + \underline{\quad} = \underline{\quad}$

2. What is 1 more than the number of ♛'s? _____

♛ ♛ ♛ ♛

3. Cross out 3 of the following ♞'s, and how many ♞'s are left? ___

♞ ♞ ♞ ♞ ♞ ♞ ♞ ♞ ♞ ♞

4. Fill in blanks

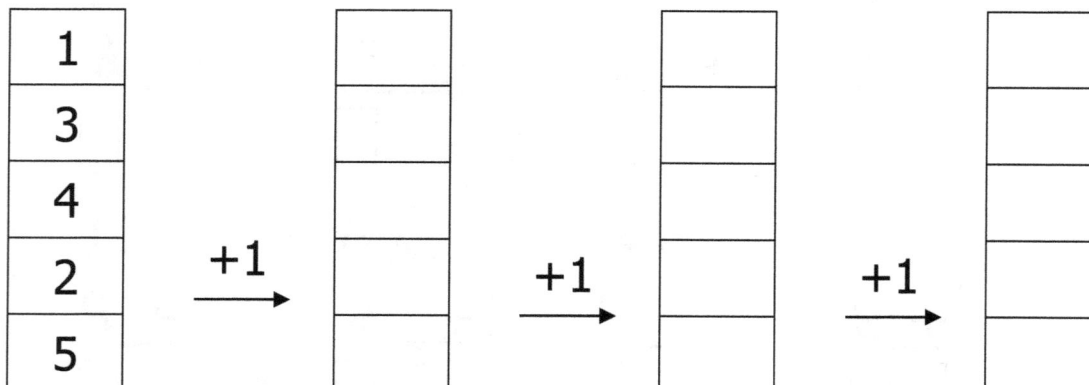

1
3
4
2
5

+1 →

+1 →

+1 →

Student's Name _____　　　　　Date _____

♔ ☖ = 0, ♕ ♙ = 1, ♗ ♝ = 3, ♘ ♞ = 3, ♖ ♜ = 5, ♛ ♕ = 9

1. Connect column 1 to match column 2.

Column 1	Column 2
♜ ♜ ♜ ♜ ♜	8
♞ ♞ ♞	6
♝ ♝ ♝ ♝ ♝	4
♛ ♛	1
♟ ♟ ♟ ♟ ♟ ♟ ♟	3
♚	5
♞ ♞ ♞ ♞	2

2. Write down the answer.

$♙ + ♙ + ♙ + ♙ = 4$

$♙ + ♙ + ♔ + ♙ + ♔ + ♙ = $ ___

$♙ + ♙ + ♙ + ♔ + ♙ = $ ___

$♙ + ♙ + ♔ + ♙ + ♙ + ♙ + ♔ = $ ___

$♙ + ♙ + ♙ + ♙ + ♔ + ♙ = $ ___

$♙ + ♔ + ♙ + ♔ + ♙ + ♙ + ♔ + ♙ = $ ___

$♔ + ♙ + ♔ + ♙ + ♔ + ♙ + ♔ + ♙ + ♙ + ♔ = $ ___

113

Student's Name _____ Date _____

1. Connect 3 columns with the same matching sum using chess piece values.

♙ + 5 ♙ (6)	4	9 + 1
♙ + 0 ♙	6	1 + 0
♙ + 9 ♙	8	3 + 5
♙ + 3 ♙	1	1 + 3
♘ + 5 ♙	10	1 + 5

2. Cross out 6 of the following ♕'s red. How many are left? ____

♕ ♕ ♕ ♕ ♕ ♕ ♕ ♕ ♕ ♕

3. Cross out 7 of the following ♘'s red. How many are left? ____

♘ ♘ ♘ ♘ ♘ ♘ ♘ ♘ ♘ ♘

4. What is 2 more than the number of ♙'s? ____

♙ ♙ ♙

5. What is 3 more than the number of ♙'s? ____

♙ ♙ ♙ ♙

6. What is 1 less than the number of ♙'s? ____ 6

♙ ♙ ♙ ♙ ♙ ♙ ♙

7. What is 2 less than the number of ♘'s? ____

♘ ♘ ♘ ♘ ♘ ♘ ♘ ♘ ♘

114

1. Find the pattern and fill in blanks.

1	☐	3	4	☐
☐	5	6	7	☐
5	4	3	☐	☐
☐	4	☐	☐	7
2	4	☐	8	☐
1	3	☐	7	☐

2. Add 1.

```
  0 9 0 8 0 1 0 1 0 1 0 1 0 1 0 1 0 1 0 1
+ 0 1 0 1 0 7 0 5 0 6 0 3 0 4 0 1 0 2 0 6 0 7
_____

  0 1 0 1 0 1 0 1 0 1 0 1 0 1 0 1 0 1 0 1 0 1
+ 0 3 0 7 0 4 0 6 0 7 0 8 0 9 0 2 0 3 0 7 0 6
_____

  0 3 0 1 0 4 0 6 0 7 0 8 0 1 0 3 0 4 0 6 0 6
+ 0 1 0 8 0 1 0 1 0 1 0 1 0 1 0 1 0 1 0 1 0 1
_____
```

Adding 2

2 + 2 = ___	2 + 3 = ___
5 + 2 = ___	7 + 2 = ___
4 + 2 = ___	8 + 2 = ___

$$..2 \\ +\ 2 \\ \square$$
$$1 \\ +\ ..2 \\ \square$$
$$..2 \\ +\ 4 \\ \square$$
$$7 \\ +\ ..2 \\ \square$$
$$5 \\ +\ ..2 \\ \square$$

$$3 \\ +\ ..2 \\ \square$$
$$..2 \\ +\ 5 \\ \square$$
$$8 \\ +\ ..2 \\ \square\square$$
$$..2 \\ +\ 7 \\ \square$$
$$6 \\ +\ ..2 \\ \square$$

$$..2 \\ +\ 5 \\ \square$$
$$..2 \\ +\ 8 \\ \square\square$$
$$6 \\ +\ ..2 \\ \square$$
$$..2 \\ +\ 6 \\ \square$$
$$4 \\ +\ ..2 \\ \square$$

Student's Name _____   Date _____

1. **Connect 3 columns with the same matching sum (chess pieces values are used).**

	3	4 + 2
♖		
⤓ + ♟ + ⤓ + ♟	5	1 + ⤓
♗	2	2 + 2
♟ + ⤓ + ♟ + ⤓ + ♟ + ♙	6	2 + ⤓
♖ + ✕	4	3 + 2
♟ + ⤓	7	6 + 2
♖ + 2	1	5 + 2
♙	8	⤓ + 0

2. **Write down the answer to adding 2.**

♙♙ + ♙♙ = 4

♙♙♙♙ + ♙♙ = ____

♙♙ + ♙♙♙ = ____

♙♙♙♙♙ + ♙♙ = ____

♙♙ + ♙♙♙♙ = ____

♙♙♙♙♙♙ + ♙♙ = ____

♙♙♙♙♙♙♙ + ♙♙ = ____

Student's Name _____   Date _____

Add 2.

You are at b2 = ☐ .	(grid)

Grid:

3	1	2	3
2	8	**2**	4
1	7	6	5
	a	b	c

Student's Name _____　　　Date _____

1. Find the pattern and fill in blanks of even numbers.

2	☐	6	8	☐
☐	12	14	16	☐
8	6	4	☐	☐
☐	4	☐	☐	10
5	7	☐	11	☐
6	8	☐	12	☐

2. Add 2.

```
  0 2 0 2 0 2 0 2 0 2 0 2 0 2 0 2 0 2 0 2 0 2 0 2
+ 0 3 0 5 0 6 0 7 0 1 0 4 0 8 0 3 0 2 0 5 0 7
```

```
  0 2 0 2 0 2 0 2 0 2 0 2 0 2 0 2 0 2 0 2 0 2 0 2
+ 0 6 0 4 0 6 0 7 0 8 0 9 0 2 0 3 0 7 0 6 0 3
```

```
  0 3 0 4 0 6 0 7 0 8 0 1 0 3 0 4 0 6 0 6 0 5
+ 0 2 0 2 0 2 0 2 0 2 0 2 0 2 0 2 0 2 0 2 0 2
```

Connect odd numbers in order starting from 1 to 81.

1	7	9	11	13	15	57	59			
3	5	23	21	19	17	55	61			
29	27	25	47	49	51	53	63			
31	37	39	45	71	69	67	65			
33	35	41	43	73	75	77	79	81	33	35	41
195	193	191	189	187	185	183	177	175	195	193	191
197	199	201	203	205	207	181	179	173	197	199	201
239	237	235	233	231	209	211	169	171	239	237	235
241	243	245	247	229	227	213	167	129	241	243	245
				223	225	215	165	131	133	123	109
				221	219	217	163	161	135	121	111
				151	153	155	157	159	137	119	113
				149	147	145	143	141	139	117	115

Student's Name _____   Date _____

1. Fill in blanks of odd numbers.

1	3	5		9		13		17	

2	4	6		10		14		18	

21	22	23		25					

10	20			50		70			

11	21	31			61				101

2. Find the answer.

♖ + 4 = 9   ♖ + 2 = _____   ♞ + 2 = _____

♝ + 5 = _____   ♛ + 6 = _____   ♚ + 6 = _____

3. Fill in each blank with an answer.

♙	2
♗	4
♖	6
7	8

1. Add 3.

2 + 3 = ___	5 + 3 = ___
6 + 3 = ___	0 + 3 = ___
4 + 3 = ___	3 + 3 = ___

2. What is 3 more than the number of ♛'s? 7

♛   ♛   ♛   ♛

3. Cross out 3 of the following ♞'s, and how many ♞'s left?

♞ ♞ ♞ ♞ ♞ ♞ ♞ ♞ ♞ ♞

4. Draw 3 more of the following ♙'s, and how many ♙'s are there? 8

♙   ♙   ♙   ♙   ♙

5. Fill in each ☐ with an answer.

Add 3.

You are at b2 = ☐.

3	3	7	1
2	2	**3**	8
1	4	6	5
	a	**b**	**c**

Student's Name _____　Date

Connect in order of counting every 3 from 3 to 372.

3	6	9	150	153	156	159	162				
54	51	12	147	144	171	168	165				
57	48	15	138	141	174	177	180				
60	45	18	135	192	189	186	183				
63	42	21	132	195	210	213	216	219	222	225	228
66	39	24	129	198	207	306	303	300	297	294	231
69	36	27	126	201	204	309	282	285	288	291	234
72	33	30	123	120	117	312	279	276	273	270	237
75	78	81	84	87	114	315	258	261	264	267	240
				90	111	318	255	252	249	246	243
				93	108	321	336	339	342	345	348
				96	105	324	333	366	363	360	351
				99	102	327	330	369	372	357	354

Student's Name _____   Date _____

Add 3.

```
  0 3 0 3 0 3 0 3 0 3 0 3 0 3 0 3 0 3 0 3 0 3
+ 0 3 0 7 0 1 0 4 0 8 0 3 0 2 0 5 0 4 0 6 0 7
```

```
  0 3 0 3 0 3 0 3 0 3 0 3 0 3 0 3 0 3 0 3 0 3
+ 0 6 0 5 0 4 0 6 0 7 0 8 0 9 0 2 0 3 0 7 0 6
```

```
  0 2 0 6 0 7 0 8 0 1 0 3 0 4 0 6 0 6 0 5 0 3
+ 0 3 0 3 0 3 0 3 0 3 0 3 0 3 0 3 0 3 0 3 0 3
```

```
  0 4 0 8 0 3 0 5 0 2 0 3 0 3 0 3 0 3 0 9 0 1
+ 0 3 0 3 0 1 0 3 0 3 0 3 0 2 0 5 0 4 0 3 0 3
```

```
  0 3 0 6 0 2 0 8 0 5 0 3 0 9 0 7 0 0 0 3 0 5
+ 0 3 0 3 0 3 0 3 0 3 0 3 0 3 0 3 0 3 0 3 0 3
```

```
  0 6 0 3 0 7 0 4 0 9 0 3 0 7 0 2 0 3 0 6 0 8
+ 0 3 0 7 0 3 0 3 0 3 0 3 0 3 0 3 0 4 0 3 0 3
```

Student's Name _____ Date _____

Adding 5

✳ ♔ = 0, ♙ = 1, ✸ ♗ = 3, ✛ ♘ = 3, ✛ ♖ = 5, ✳ ♕ = 9

Note 9 + d = 9 + 1 + (d−1), 9 + 5 = 9 + 1 + 4, so the answer is 14.

✛ + ⋎　　5 + 1 = 6	✛ + ✛　　5 + 3 = ___
✛ + ✛　　___ + ___ = ___	✛ + ✳　　___ + ___ = ___
✛ + ✳　　___ + ___ = ___	✳ + ✛　　___ + ___ = ___

✛	1	2	✳	✛
+ ✳	+ ✛	+ ✛	+ ✛	+ ✸
□	□	□	□	□

✛	✛	✛	✛	✛
+ 4	+ ✸	+ ⋎	+ ✳	+ ✛
□	□	□	□	□

✛	✛	✛	4	✳
+ ✛	+ ✛	+ 3	+ ✛	+ ✛
□	□□	□	□	□

Student's Name _____   Date _____

✳ ♔ = 0, ♕ ♙ = 1, ✖ ♝ = 3, ✜ ♞ = 3, ✛ ♖ = 5, ❋ ♛ = 9

1. Calculate d + ? = 5 = d + ?, Adding to 5.

$$✳ + \boxed{} = \leftrightarrow = \leftrightarrow + \boxed{}$$

$$\downarrow + \boxed{} = 5 = 4 + \boxed{}$$

$$2 + \boxed{} = \leftrightarrow = ✜ + \boxed{}$$

$$⤢ + \boxed{} = 5 = 2 + \boxed{}$$

$$4 + \boxed{} = \leftrightarrow = \downarrow + \boxed{}$$

$$\leftrightarrow + \boxed{} = 5 = ✳ + \boxed{}$$

2. Add 3.

4
6
↔
✜
7
↓
⤢

$+$ ✜

Ho Math Chess　何数棋谜　低年级棋谜式数学

© 2007 – 2020　　FRANK HO, AMANDA HO　　All rights reserved. www.homathchess.com

Student's Name _____　Date _____

Connect in order of every 5 from 5 to 620.

5	10	15	70	75	90	95	100				
30	25	20	65	80	85	110	105				
35	50	55	60	155	150	115	120				
40	45	170	165	160	145	140	125				
185	180	175	210	215	220	135	130	455	460	465	470
190	195	200	205	230	225	430	435	450	485	480	475
305	300	245	240	235	420	425	440	445	490	525	530
310	295	250	375	380	415	410	405	500	495	520	535
315	290	255	370	385	390	395	400	505	510	515	540
320	285	260	365					580	575	570	545
325	280	265	360					585	590	565	550
330	275	270	355					600	595	560	555
335	340	345	350					605	610	615	620

128

Student's Name _____   Date _____

Connect in order of every 10 from 10 to 1400.

10	20	30	40	50	560	570	580	590	600	710	720
120	110	100	70	60	550	640	630	620	610	700	730
130	140	90	80	530	540	650	660	670	680	690	740
160	150	200	210	520	510	500	490	480	830	820	750
170	180	190	220					470	840	810	760
300	290	280	230					460	850	800	770
310	320	270	240					450	860	790	780
340	330	260	250					440	870	880	890
350	360	370	380	390	400	410	420	430	920	910	900
1100	1090	1080	1070	1060	1050	1040	1030	1020	930	1360	1370
1110	1220	1230	1240	1250	1260	990	1000	1010	940	1350	1380
1120	1210	1200	1190	1180	1270	980	970	960	950	1340	1390
1130	1140	1150	1160	1170	1280	1290	1300	1310	1320	1330	1400

Student's Name _____　　　　Date _____

9 plus d

♛ = 0, ♟ = 1, ♝ = 3, ♞ = 3, ♜ = 5, ♕ = 9

$9 + d = 9 + 1 + (d - 1)$

1. Fill in blanks of 9 + d.

❋. ↓ . . 9 + 1 = 10	❋. ♘ . . . 9 + 3 = ___
❋. ⚓ . . . ___ + ___ = ___	⚓. ❋ . . . ___ + ___ = ___
❋. ✳ . . . ___ + ___ = ___	⚰ . ❋ . . . ___ + ___ = ___

2. Mixed computations using chess pieces values

⚰ + 2 = 5　　　　9 + ♞ = ____　　　　♞ + 3 = ____

⚓ + 5 = ____　　7 + ⚓ = ____　　❋ + 7 = ____

❋ + 4 = ____　　8 + ✳ = ____　　✳ + 4 = ____

3. Fill in each ☐ with an answer.

11　13　15　♕　♟　7　♖　☐	

Student's Name _____   Date _____

Doubling

♔ = 0, ♙ = 1, ♗ = 3, ♘ = 3, ♖ = 5, ♕ = 9

1. Find the number of each shape represents.

(1)   If □ + □ = 8, then □ = _____

(2)   If ○ + ○ = 12, then ○ = _____

(3)   If ◇ + ◇ = 6, then ◇ = _____

(4)   If ★ + ★ = 14, then ★ = _____

(5)   If ◇ + ◇ = 16, then ◇ = _____

(6)   If * + * = 10, then * = _____

(7)   If ⚘ + ⚘ = 18, then ⚘ = _____

(8)   If ✳ + ✳ = 20, then ✳ = _____

2. Fill in the following table.

	9	
6	8	?
	17	

	17	
?	9	4
	26	

	17	
?	9	3
	8	

3. Fill in the following ☐ with a number.

♘ , 6, ♕ , 12, ☐

1. Fill in blanks using double of chess values.

♔ = 0, ♙ = 1, ♗ = 3, ♘ = 3, ♖ = 5, ♕ = 9

(♖ + ♙)+(♖ + ♙)= 6 + 6 = 12	(♖ + ♖)+(♖ + ♖) 10 + 10 = ___
(♗ + ♗)+(♗ + ♗) ___ + ___ = ___	♕ + ♕ ___ + ___ = ___
(♖ + ♘)+(♖ + ♘) ___ + ___ = ___	(♘ + ♘)+(♘ + ♘) ___ + ___ = ___

2. Circle the largest number in each of the following number sequences.

1	6	4	10	5			4	8	6	12	7
16	3	9	21	17			3	9	17	42	15
15	31	32	51	46			27	35	54	37	18
42	18	25	49	17			51	28	16	9	11
73	81	24	47	22			8	38	48	21	62
65	32	57	23	66			11	30	91	19	28

Student's Name _____　　　　　　　Date _____

Doubling

♛ ♚ = 0, ♖ ♙ = 1, ✖ ♝ = 3, ♘ = 3, ♖ = 5, ♛ = 9

1. Doubling

1 + ⬇ = 2

3 + ✚ = _____

5 + ↔ = _____

7 + 7 = _____

✦ + 9 = _____

2 + 2 = _____

4 + 4 = _____

6 + 6 = _____

8 + 8 = _____

10 + 10 = _____

2. Fill in blanks using chess values.

♙ + ♙ 1 + 1 = 2	♞ + ♞ 3 + 3 = ___
♙♙ + ♙♙ ___ + ___ = ___	♖ + ♖ ___ + ___ = ___
♜ + ♜ ___ + ___ = ___	♔ + ♔ ___ + ___ = ___
♕ + ♕ ___ + ___ = _____	♖ + ♖ ___ + ___ = _____
♝ + ♝ ___ + ___ = _____	♙ + ♙ ___ + ___ = _____

Student's Name _____ Date _____

Doubles

Row 1

Box 1:
$$\lceil + \quad 6 \quad + \rceil$$
$$6 \quad + \quad 7$$
$$\square \qquad 6 \qquad \square$$
$$+$$
$$1$$
$$\square$$

Box 2:
$$\lceil + \quad 6 \quad + \rceil$$
$$6 \quad + \quad 8$$
$$\square \qquad 6 \qquad \square$$
$$+$$
$$2$$
$$\square$$

Box 3:
$$\lceil + \quad 5 \quad + \rceil$$
$$5 \quad + \quad 6$$
$$\square \qquad 5 \qquad \square$$
$$+$$
$$1$$
$$\square$$

Row 2

Box 4:
$$\lceil + \quad 5 \quad + \rceil$$
$$5 \quad + \quad 7$$
$$\square \qquad 5 \qquad \square$$
$$+$$
$$2$$
$$\square$$

Box 5:
$$\lceil + \quad 7 \quad + \rceil$$
$$7 \quad + \quad 8$$
$$\square \qquad 7 \qquad \square$$
$$+$$
$$1$$
$$\square$$

Box 6:
$$\lceil + \quad 7 \quad + \rceil$$
$$7 \quad + \quad 9$$
$$\square \qquad 7 \qquad \square$$
$$+$$
$$2$$
$$\square$$

Row 3

Box 7:
$$\lceil + \quad 4 \quad + \rceil$$
$$4 \quad + \quad 6$$
$$\square \qquad 4 \qquad \square$$
$$+$$
$$2$$
$$\square$$

Box 8:
$$\lceil + \quad 4 \quad + \rceil$$
$$4 \quad + \quad 5$$
$$\square \qquad 4 \qquad \square$$
$$+$$
$$1$$
$$\square$$

Box 9:
$$\lceil + \quad 5 \quad + \rceil$$
$$5 \quad + \quad 6$$
$$\square \qquad 5 \qquad \square$$
$$+$$
$$1$$
$$\square$$

Doubling using chess pieces values

Student's Name _____   Date _____

1. Double the following value of chess pieces.

♙	2	♙ + ♞	8	♙ + ♙	
♞		♖		♛	
♗ + ♗		♔ + ♞	6	7	
4 + ♗		♔ + 2		8	
♖ + ♖		6		5	
♛		♗ + ♙		♝	
♖ + ♙		♞ + ♖		♞ + ♞	
♔ + ♖		7 + ♙		4 + ♞	

2. Connect each value of double chess pieces.

♞ + ♞ (6)	10	2♔
♛ + ♛	6	2♖
♔ + ♔	2	2♞
♖ + ♖	0	2♙
♙ + ♙	18	2♛

Student's Name _____   Date _____

1 Calculate using double plus 1.

$3 + 3 = 6$ $3 + 4 = 3 + 3 + 1 = 6 + 1 = 7$

$2 + 2 =$ _____ $2 + 3 =$ _____

$5 + 5 =$ _____ $5 + 6 =$ _____

$7 + 7 =$ _____ $7 + 8 =$ _____

$9 + 9 =$ _____ $9 + 10 =$ _____ (add directly.)

$4 + 4 =$ _____ $4 + 5 =$ _____

$6 + 6 =$ _____ $6 + 7 =$ _____

$8 + 8 =$ _____ $8 + 9 =$ _____

$7 + 7 =$ _____ $7 + 8 =$ _____

2. Write a number sentence using chess values.

♖ + ♖   ♖ + 6

$5 + 5 = 10$ ___ + ___ = ___

♕ + ♕   ♕ + 10

___ + ___ = ___   ___ + ___ = ___

♘ + ♘   ♘ + 4

___ + ___ = ___   ___ + ___ = ___

♔ + ♔   ♔ + ♙

___ + ___ = ___   ___ + ___ = ___

Student's Name _____   Date _____

Doubling

1. Find the pattern and fill in blanks.

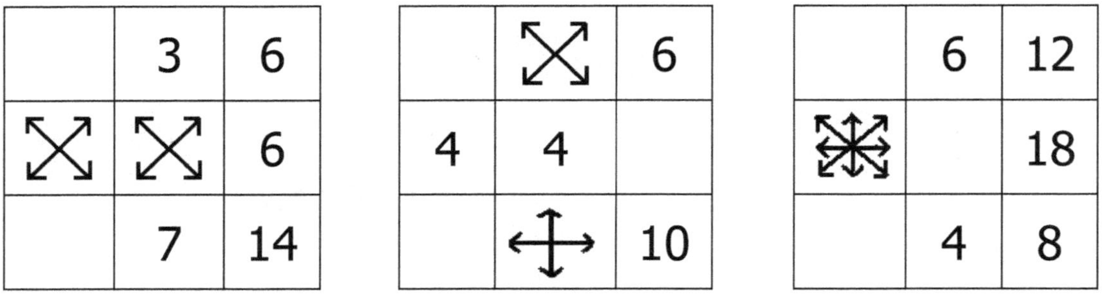

2. Fill in blanks.

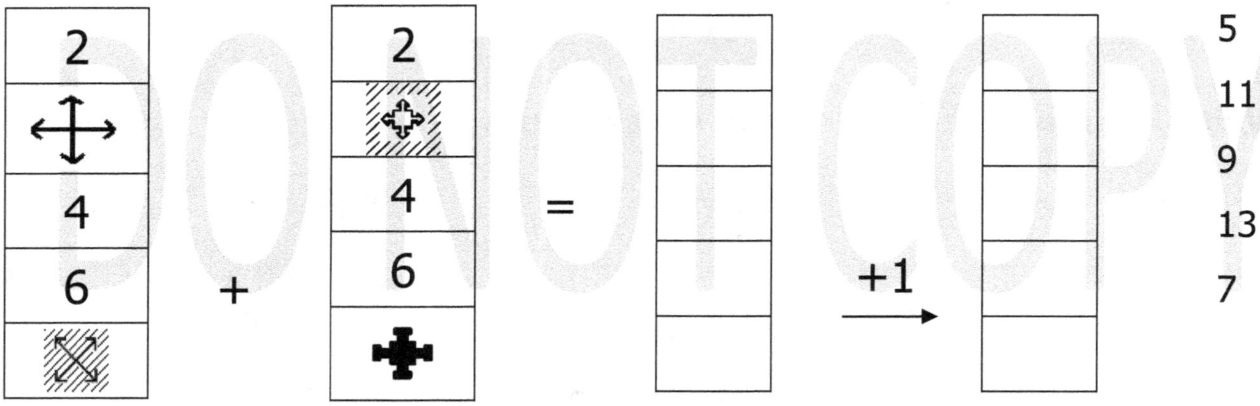

3. Find the answer to the question mark in the following diagram.

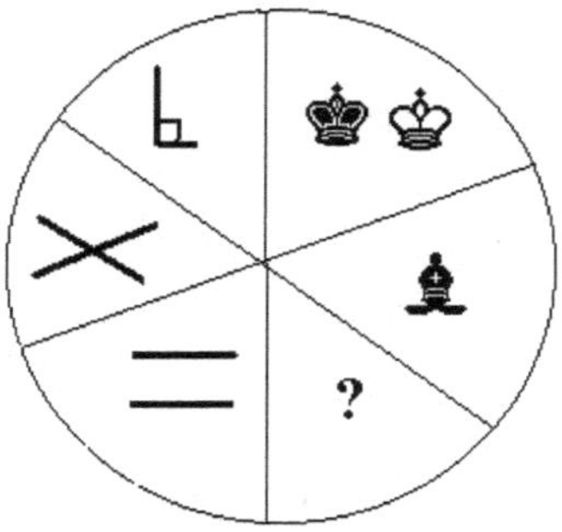

Student's Name _____   Date _____

Find the missing number of doubles.

Chicks	1	2	3	4	5	6	7	8	9	10
Legs										

Rabbits	1	3	7	2	8	4	6	10	5	9
Ears										

Dogs	5	7	2	6	9	3	1	8	4	10
Eyes										

Boxes	6	3	7	8	1	10	20	5	9	4
Cookies										

Boys	2	5	1	8	3	9	6	4	5	7
Arms										

Legs	2	8	6	4	10	12	18	14	16	20
Birds										

Student's Name _____ Date _____

Double and double + 1

```
  0 5 0 7 0 7 0 6 0 6 0 4 0 4 0 2 0 2 0 7 0 7
+ 0 5 0 8 0 7 0 7 0 6 0 5 0 4 0 3 0 2 0 8 0 7
```

```
  0 3 0 4 0 4 0 5 0 5 0 7 0 7 0 3 0 3 0 6 0 6
+ 0 3 0 5 0 4 0 6 0 5 0 8 0 7 0 4 0 3 0 7 0 6
```

```
  0 6 0 4 0 4 0 7 0 7 0 7 0 7 0 4 0 4 0 6 0 6
+ 0 6 0 5 0 4 0 8 0 7 0 8 0 7 0 5 0 4 0 7 0 6
```

```
  0 3 0 6 0 6 0 8 0 8 0 6 0 6 0 2 0 2 0 5 0 5
+ 0 3 0 7 0 6 0 9 0 8 0 7 0 6 0 3 0 2 0 6 0 5
```

```
  0 9 0 5 0 4 0 6 0 5 0 8 0 7 0 9 0 8 0 4 0 4
+ 0 9 0 4 0 4 0 5 0 5 0 7 0 7 0 8 0 8 0 3 0 4
```

```
  0 6 0 7 0 4 0 9 0 7 0 9 0 6 0 6 0 4 0 4 0 7
+ 0 5 0 8 0 5 0 8 0 8 0 8 0 5 0 5 0 3 0 5 0 6
```

Student's Name   Date

1. **Add using doubles plus 1.**

$$6 + 1 = \boxed{}$$

$$+ \quad 6 + 0 = \boxed{} \quad +$$

$$\boxed{} \qquad \boxed{}$$

If 6 + 6 = $\boxed{}$, then 7 + 6 must be $\boxed{}$.

If 6 + 6 = $\boxed{}$, then 6 + 7 must be $\boxed{}$.

2. If 6 plus 6 is $\boxed{}$, what is 7 plus 6?

Answer: 7 + 6 = $\boxed{}$

3. If 6 plus 6 is $\boxed{}$, what is 6 plus 8?

Answer: 6 + 8 = $\boxed{}$

4. At the end of a chess game, White had a rook left. Black has 3 more points than White. How many points does Black have?

Student's Name _____　　Date _____

1. Calculate using double plus 2.

$3 + 3 = 6$　　　　　　　　　　　$3 + 5 = 3 + 3 + 2 = 6 + 2 = 8$

$4 + 4 = $ ____　　　　　　　　　　$4 + 6 = $ ____

$5 + 5 = $ ____　　　　　　　　　　$5 + 7 = $ ____

$7 + 7 = $ ____　　　　　　　　　　$7 + 9 = $ ____

$6 + 6 = $ ____　　　　　　　　　　$6 + 8 = $ ____

2. Fill in blanks.

♖ + ♖
$5 + 5 = 10$

♖ + 7
____ + ____ = ____

♕ + ♕
____ + ____ = ____

♕ + 11
____ + ____ = ____

♘ + ♘
____ + ____ = ____

♘ + 5
____ + ____ = ____

♔ + ♔
____ + ____ = ____

♔ + 2
____ + ____ = ____

1. Find the pattern and fill in blanks.

4	5	?
3	3	6
?	8	?

?	7	12
5	5	?
10	?	?

?	6	?
4	?	8
?	10	19

2. Find the pattern and fill in blanks.

1. $9 + d = 9 + 1 + (d−1)$.

$$7 + 2 = \boxed{}9$$

$$+ \qquad 7 + 0 = \boxed{}7 \qquad +$$

$$14\boxed{} \qquad\qquad \boxed{}16$$

If $10 + 6 = \boxed{}16$, then $9 + 7$ $(9+1+6)$ must be $\boxed{}16$.

If $6 + 10 = \boxed{}16$, then $7 + 9$ must be $\boxed{}16$.

2. If 10 plus 6 is $\boxed{}16$, **what is 9 plus 7?**

Answer: $9 + 7 = \boxed{}16$

3. If 6 plus 10 is $\boxed{}16$, **what is 7 plus 9?**

Answer: $7 + 9 = \boxed{}16$

4. At the end of a game, White and Black have 12 points left. White and Black have the same number of points. How many points does each have?

Student's Name _____　Date _____

Double and 9 + d

```
  0 7 0 7 0 6 0 6 0 4 0 4 0 2 0 2 0 7 0 7 0 3
+ 0 9 0 7 0 8 0 6 0 6 0 4 0 4 0 2 0 9 0 7 0 3
```

```
  0 6 0 4 0 7 0 5 0 8 0 8 0 9 0 7 0 6 0 4 0 8
+ 0 4 0 4 0 5 0 5 0 6 0 8 0 7 0 7 0 4 0 4 0 8
```

```
  0 4 0 4 0 6 0 6 0 6 0 8 0 5 0 5 0 7 0 7 0 6
+ 0 6 0 4 0 4 0 6 0 6 0 8 0 7 0 5 0 9 0 7 0 6
```

```
  0 3 0 6 0 6 0 5 0 5 0 5 0 3 0 6 0 6 0 5 0 5
+ 0 3 0 8 0 6 0 8 0 5 0 3 0 3 0 8 0 6 0 7 0 5
```

```
  0 8 0 6 0 4 0 7 0 5 0 9 0 7 0 6 0 6 0 7 0 7
+ 0 8 0 4 0 4 0 5 0 5 0 7 0 7 0 8 0 6 0 9 0 7
```

```
  0 4 0 5 0 5 0 7 0 7 0 5 0 5 0 3 0 3 0 6 0 6
+ 0 4 0 7 0 5 0 9 0 7 0 7 0 5 0 5 0 3 0 8 0 6
```

Doubles

1. Fill in the following ☐ with a number.

6 $+\ 7$ ☐ $- 7 =$ ☐	7 $+\ 6$ ☐ $- 6 =$ ☐
7 $+\ 7$ ☐ $- 7 =$ ☐	7 $+\ 7$ ☐ $- 7 =$ ☐
7 $+\ 8$ ☐ $- 8 =$ ☐	8 $+\ 7$ ☐ $- 7 =$ ☐
7 $+\ 9$ ☐ $- 9 =$ ☐	9 $+\ 7$ ☐ $- 7 =$ ☐

2. Fill in blanks.

6		5		4		3
3						
5	-1		-1		$-1 =$	
9						
7						
8						

Adding to 10

♛	1	2	8	♞
+ 1	+ ♛	+ 8	+ 2	+ 7

7	6	4	♜	♝
+ ♝	+ 4	+ 6	+ 5	+ 7

7	2	8	1	♛
+ ♞	+ 8	+ 2	+ ♛	+ 1

♝	☐	♜	♙	4
+ ☐	+ 5	+ ☐	+ ☐	+ ☐
1 0	1 0	1 0	1 0	1 0

♛	♞	♙	♜	♛
+ ☐	+ ☐	+ ☐	+ ☐	+ ☐
1 0	1 0	1 0	1 0	1 0

Student's Name _____　　　　　Date _____

Adding to 10

1. Circle the two numbers next to each other in moves as indicated by the chess piece on the chessboard that makes the sum of 10. Find as many answers as you can

6	5	2	3	5
3	8	8	5	0
9	1	♛	2	8
3	4	6	3	2
6	7	4	8	7

6	5	2	4	5
5	8	8	5	6
9	1	♞	2	8
3	4	6	3	2
6	7	4	8	7

2. Circle the two numbers next to each other that make the sum of 9. Find as many answers you can.

1	7	3	1	6
2	5	7	4	9
7	8	6	5	8
2	5	9	1	5
6	8	3	6	4

2	9	4	6	3
7	8	3	6	5
3	1	9	8	4
8	2	5	9	3
2	7	1	7	6

Student's Name _____   Date _____

1. Adding to 10

Fill in the following ☐ with a chess piece.	Fill in the following ☐ with a number.	
♙	9	10
♖	☐	10
♕	☐	10
☐ or ☐	7	10

2. At the end of a chess game, White and Black total has 10 points left. White has two more points than Black. How many points does White have?

Student's Name _____   Date _____

1. Adding to 10

♛ + ♟	♜ + ♜
9 + 1 = 10	5 + 5 = ___
♝ +7 ___ + ___ = ___	♚ +10 ___ + ___ = ___
(♜ + ♞) +2 ___ + ___ = ___	(♞ + ♞) + (♞ + ♟) ___ + ___ = ___

2. Fill in blanks.

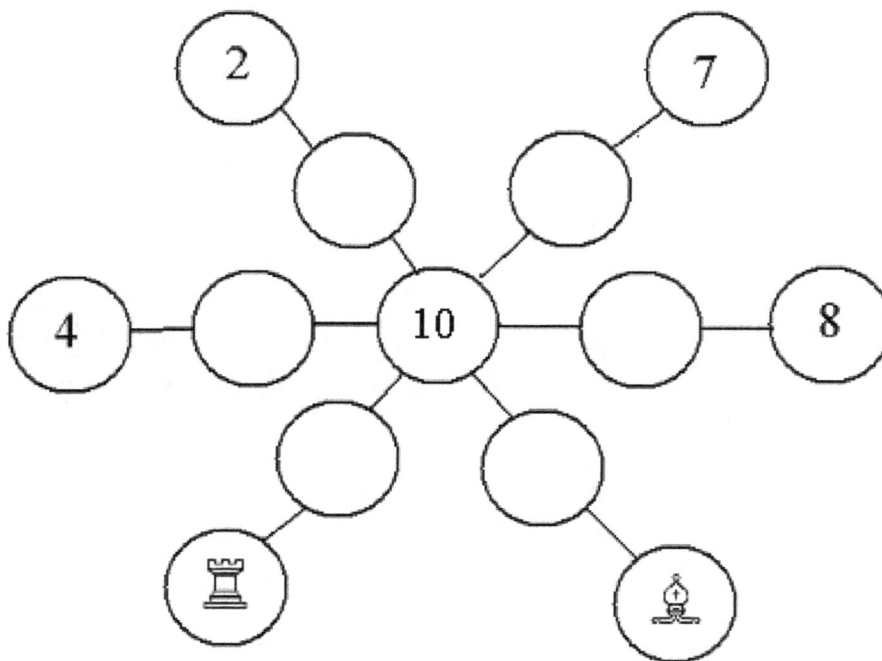

Student's Name _____　Date

1. Compute.

$$10 - 4 = \square \qquad 10 - 8 = \square \qquad 10 - 7 = \square \qquad 10 - 3 = \square \qquad 10 - 6 = \square$$

$$10 - 6 = \square \qquad 10 - 1 = \square \qquad 10 - 4 = \square \qquad 10 - 5 = \square \qquad 10 - 2 = \square$$

$$10 - 7 = \square \qquad 10 - 9 = \square \qquad 10 - 2 = \square \qquad 10 - 7 = \square \qquad 10 - 1 = \square$$

2. Compute.

1.	10 − 1 =	2.	10 − 6 =
3.	10 − 5 =	4.	10 − 3 =
5.	10 − 3 =	6.	10 − 8 =
7.	10 − 7 =	8.	10 − 9 =
9.	10 − 5 =	10.	10 − 5 =
11.	10 − 8 =	12.	10 − 1 =
13.	10 − 9 =	14.	10 − 6 =
15.	10 − 2 =	16.	10 − 7 =

Student's Name _____   Date _____

Adding 10

You are at b2 = ☐.

3	1	8	9
2	7	**10**	5
1	4	6	3
	a	b	c

$10 - $ $=$

$10 - $ $=$

$10 - $ $=$

$10 - $ $=$

$10 - $ $=$

$10 - $ $=$

$10 - $ $=$

$10 - $ $=$

$10 - $ $=$

$10 - $ $=$

$10 - $ $=$

$10 - $ $=$

$10 - $ $=$

$10 - $ $=$

$10 - $ $=$

$10 - $ $=$

Student's Name _____   Date _____

Adding to 10 plus 1

⌐+ 5 +⌐
5 + 6
□   □
   5
   +
   1
   □

⌐+ 4 +⌐
6 + 7
□   □
   6
   +
   1
   □

⌐+ 3 +⌐
7 + 8
□   □
   7
   +
   1
   □

⌐+ 6 +⌐
4 + 5
□   □
   4
   +
   1
   □

⌐+ 7 +⌐
3 + 4
□   □
   3
   +
   1
   □

⌐+ 8 +⌐
2 + 3
□   □
   2
   +
   1
   □

⌐+ 9 +⌐
1 + 2
□   □
   1
   +
   1
   □

⌐+ 2 +⌐
8 + 9
□   □
   8
   +
   1
   □

⌐+ 1 +⌐
9 + 10
□   □
   9
   +
   1
   □

Adding to 10.

```
  0 1 0 2 0 3 0 5 0 4 0 7 0 6 0 9 0 8 0 4 0 3
+ 0 9 0 8 0 7 0 5 0 6 0 3 0 4 0 1 0 2 0 6 0 7
```

```
  0 7 0 3 0 9 0 4 0 3 0 2 0 1 0 8 0 7 0 3 0 4
+ 0 3 0 7 0 1 0 6 0 7 0 8 0 9 0 2 0 3 0 7 0 6
```

```
  0 7 0 3 0 6 0 4 0 3 0 2 0 3 0 7 0 6 0 4 0 4
+ 0 3 0 7 0 4 0 6 0 7 0 8 0 7 0 3 0 4 0 6 0 6
```

```
  0 7 0 3 0 6 0 4 0 3 0 2 0 3 0 7 0 6 0 4 0 4
+ 0 3 0 7 0 4 0 6 0 7 0 8 0 7 0 3 0 4 0 6 0 6
```

```
  0   0   0   0   0   0   0   0   0   0   0
+ 0 3 0 7 0 4 0 6 0 7 0 7 0 7 0 3 0 8 0 3 0 4
  1 0 1 0 1 0 1 0 1 0 1 0 1 0 1 0 1 0 1 0 1 0
```

```
  0 6 0 7 0 6 0 8 0 6 0 7 0 6 0 6 0 6 0 6 0 7
+ 0   0   0   0   0   0   0   0   0   0   0
  1 0 1 0 1 0 1 0 1 0 1 0 1 0 1 0 1 0 1 0 1 0
```

Student's Name _____   Date _____

Adding to 10 plus 2

⌐+ 2 +⌐	⌐+ 4 +⌐	⌐+ 3 +⌐
8 + 10	6 + 8	7 + 9
□ □	□ □	□ □
8	6	7
+	+	+
2	2	2
□	□	□

⌐+ 6 +⌐	⌐+ 7 +⌐	⌐+ 8 +⌐
4 + 6	3 + 5	2 + 4
□ □	□ □	□ □
4	3	2
+	+	+
2	2	2
□	□	□

⌐+ 9 +⌐	⌐+ 2 +⌐	⌐+ 5 +⌐
1 + 3	8 + 10	5 + 7
□ □	□ □	□ □
1	8	5
+	+	+
2	2	2
□	□	□

Student's Name _____   Date _____

1. Compute.

$$
\begin{array}{r} 10 \\ -\ 4 \\ \hline \square \end{array}
\qquad
\begin{array}{r} 10 \\ -\ 8 \\ \hline \square \end{array}
\qquad
\begin{array}{r} 10 \\ -\ 7 \\ \hline \square \end{array}
\qquad
\begin{array}{r} 10 \\ -\ 3 \\ \hline \square \end{array}
\qquad
\begin{array}{r} 10 \\ -\ 6 \\ \hline \square \end{array}
$$

$$
\begin{array}{r} 10 \\ -\ 6 \\ \hline \square \end{array}
\qquad
\begin{array}{r} 10 \\ -\ 1 \\ \hline \square \end{array}
\qquad
\begin{array}{r} 10 \\ -\ 4 \\ \hline \square \end{array}
\qquad
\begin{array}{r} 10 \\ -\ 5 \\ \hline \square \end{array}
\qquad
\begin{array}{r} 10 \\ -\ 2 \\ \hline \square \end{array}
$$

$$
\begin{array}{r} 10 \\ -\ 7 \\ \hline \square \end{array}
\qquad
\begin{array}{r} 10 \\ -\ 9 \\ \hline \square \end{array}
\qquad
\begin{array}{r} 10 \\ -\ 2 \\ \hline \square \end{array}
\qquad
\begin{array}{r} 10 \\ -\ 7 \\ \hline \square \end{array}
\qquad
\begin{array}{r} 10 \\ -\ 1 \\ \hline \square \end{array}
$$

2. Compute.

1. $10 - 1 =$	2. $10 - 6 =$
3. $10 - 5 =$	4. $10 - 3 =$
5. $10 - 3 =$	6. $10 - 8 =$
7. $10 - 7 =$	8. $10 - 9 =$
9. $10 - 5 =$	10. $10 - 5 =$
11. $10 - 8 =$	12. $10 - 1 =$
13. $10 - 9 =$	14. $10 - 6 =$
15. $10 - 2 =$	16. $10 - 7 =$

The reverse of adding 10

```
  1 0 1 0 1 0 1 0 1 0 1 0 1 0 1 0 1 0 1 0 1 0
− 0 2 0 4 0 6 0 3 0 7 0 3 0 9 0 5 0 3 0 8 0 5
```

```
  1 0 1 0 1 0 1 0 1 0 1 0 1 0 1 0 1 0 1 0 1 0
− 0 7 0 3 0 6 0 4 0 3 0 5 0 3 0 5 0 7 0 4 0 3
```

```
  1 0 1 0 1 0 1 0 1 0 1 0 1 0 1 0 1 0 1 0 1 0
− 0 5 0 4 0 2 0 4 0 6 0 8 0 5 0 7 0 8 0 4 0 7
```

```
  1 0 1 0 1 0 1 0 1 0 1 0 1 0 1 0 1 0 1 0 1 0
− 0 3 0 6 0 7 0 5 0 8 0 4 0 9 0 4 0 6 0 7 0 2
```

```
  1 0 1 0 1 0 1 0 1 0 1 0 1 0 1 0 1 0 1 0 1 0
− 0 4 0 6 0 3 0 6 0 7 0 9 0 7 0 5 0 6 0 4 0 7
```

```
  1 0 1 0 1 0 1 0 1 0 1 0 1 0 1 0 1 0 1 0 1 0
− 0 6 0 8 0 7 0 4 0 7 0 5 0 1 0 4 0 4 0 5 0 3
```

Student's Name _____　Date _____

Adding to 10 and doubling

```
  0 1 0 1 0 3 0 2 0 4 0 6 0 5 0 7 0 8 0 2 0 1
+ 0 9 0 8 0 7 0 5 0 6 0 3 0 4 0 1 0 2 0 6 0 7
```

```
  0 2 0 3 0 2 0 1 0 3 0 2 0 1 0 5 0 6 0 2 0 4
+ 0 3 0 7 0 4 0 6 0 7 0 8 0 9 0 2 0 3 0 7 0 6
```

```
  0 3 0 2 0 1 0 4 0 2 0 1 0 2 0 5 0 6 0 3 0 2
+ 0 3 0 7 0 4 0 6 0 7 0 8 0 7 0 3 0 4 0 6 0 6
```

```
  0 1 0 6 0 2 0 1 0 2 0 8 0 5 0 1 0 7 0 4 0 3
+ 0 7 0 4 0 6 0 7 0 8 0 1 0 4 0 7 0 3 0 4 0 6
```

```
  0 5 0 6 0 3 0 2 0 1 0 5 0 6 0 1 0 7 0 2 0 7
+ 0 4 0 4 0 6 0 7 0 7 0 5 0 3 0 7 0 3 0 8 0 3
```

```
  0 3 0 7 0 6 0 1 0 6 0 3 0 2 0 2 0 2 0 6 0 5
+ 0 5 0 3 0 3 0 7 0 4 0 6 0 7 0 8 0 7 0 3 0 4
```

Ho Math Chess 何数棋谜 低年级棋谜式数学

Student's Name _____ Date _____

Adding to 10

You are at b2 = ☐.

3	1	8	9
2	7	**10**	5
1	4	6	3
	a	b	c

10 – ☐ =	10 – ☐ =
10 – ☐ =	10 – ☐ =
10 – ☐ =	10 – ☐ =
10 – ☐ =	10 – ☐ =
10 – ☐ =	10 – ☐ =
10 – ☐ =	10 – ☐ =
10 – ☐ =	10 – ☐ =
10 – ☐ =	10 – ☐ =

Student's Name _____   Date _____

Subtracting

Counting

1. Count back by 1.

10	9	8						

20		18		16		14		12	

30			26				

15		13					

28				22			

12			9				

2. Fill in blanks.

5
9
3
7
4

− 1 =

5

− 1 =

4

+ 2 =

6

160

Student's Name _____　　Date _____

Subtraction 1

1. Compute.

$$4 - 1 = \boxed{} \quad 7 - 1 = \boxed{} \quad 2 - 1 = \boxed{} \quad 6 - 1 = \boxed{} \quad 3 - 1 = \boxed{}$$

$$4 - 1 = \boxed{} \quad 3 - 1 = \boxed{} \quad 7 - 1 = \boxed{} \quad 5 - 1 = \boxed{} \quad 9 - 1 = \boxed{}$$

$$8 - 1 = \boxed{} \quad 4 - 1 = \boxed{} \quad 3 - 1 = \boxed{} \quad 6 - 1 = \boxed{} \quad 5 - 1 = \boxed{}$$

2. Compute.

1.　$4 - 1 =$	2.　$1 - 1 =$
3.　$8 - 1 =$	4.　$7 - 1 =$
5.　$4 - 1 =$	6.　$4 - 1 =$
7.　$7 - 1 =$	8.　$9 - 1 =$
9.　$2 - 1 =$	10.　$5 - 1 =$
11.　$6 - 1 =$	12.　$8 - 1 =$
13.　$4 - 1 =$	14.　$2 - 1 =$
15.　$7 - 1 =$	16.　$5 - 1 =$

Student's Name _____   Date _____

Subtraction 1

Compute.

You are at b2 = ☐.

3	6	2	9
2	4	**1**	5
1	1	7	3
	a	b	c

Student's Name _____   Date _____

Subtraction 1

```
  2 5 3 2 5 4 3 3 4 1 2 3 6 2 7 8 3 4 1 7 3
- 1 1 1 1 1 1 1 1 1 1 1 1 1 1 1 1 1 1 1 1 1
```

```
  3 6 4 5 7 3 2 1 3 4 5 3 5 1 3 6 3 7 3 8 2
- 1 1 1 1 1 1 1 1 1 1 1 1 1 1 1 1 1 1 1 1 1
```

```
  5 2 7 1 9 4 3 6 5 4 7 5 2 7 2 4 8 3 9 3 1
- 1 1 1 1 1 1 1 1 1 1 1 1 1 1 1 1 1 1 1 1 1
```

```
  6 4 5 8 2 7 3 9 7 4 6 3 8 5 9 2 4 8 5 3 9
- 1 1 1 1 1 1 1 1 1 1 1 1 1 1 1 1 1 1 1 1 1
```

```
  5 7 3 9 4 8 5 3 2 9 7 1 4 6 3 8 4 2 4 9 6
- 1 1 1 1 1 1 1 1 1 1 1 1 1 1 1 1 1 1 1 1 1
```

```
  3 2 8 5 4 7 6 3 9 6 1 5 8 5 3 1 4 6 3 9 4
- 1 1 1 1 1 1 1 1 1 1 1 1 1 1 1 1 1 1 1 1 1
```

Student's Name _____   Date _____

Complete in backward (subtractions).

```
  0 2 0 7 0 5 0 2 0 7 0 5 0 6 0 4 0 7 0 6 0 4
+ 0   0   0   0   0   0   0   0   0   0   0 6
  1 1 1 3 1 2 1 0 1 4 1 1 1 4 0 5 1 6 1 3 1 0
```

```
  0 6 0 8 0 2 0 7 0 5 0 6 0 3 0 7 0 4 0 6 0 9
+ 0   0   0   0   0   0   0   0   0   0   0 7
  1 2 1 4 0 6 1 3 1 1 1 3 0 7 1 2 1 0 1 1 1 6
```

```
  0 3 0 7 0 5 0 8 0 4 0 8 0 9 0 5 0 7 0 6 0 7
+ 0   0   0   0   0   0   0   0   0   0   0 6
  0 8 1 2 1 1 1 6 1 0 1 4 1 6 1 2 1 6 1 4 1 3
```

```
  0 8 0 9 0 4 0 3 0 7 0 8 0 7 0 5 0 9 0 4 0 8
+ 0   0   0   0   0   0   0   0   0   0   0 6
  1 6 1 6 1 0 0 6 1 2 1 6 1 2 1 1 1 7 0 9 1 4
```

```
  0 9 0 7 0 6 0 3 0 7 0 5 0 3 0 8 0 9 0 4 0 3
+ 0   0   0   0   0   0   0   0   0   0   0 6
  1 8 1 3 1 2 0 7 1 3 1 1 0 8 1 5 1 7 0 9 0 9
```

Student's Name _____   Date _____

Subtractions

$2 + 7 = \square$ Think 7 + 2	$9 - 2 = \square$	$9 - 7 = \square$
$6 + 1 = \square$	$7 - 1 = \square$	$7 - 6 = \square$
$3 + 5 = \square$ Think 5 + 3	$8 - 3 = \square$	$8 - 5 = \square$
$6 + 3 = \square$	$9 - 3 = \square$	$9 - 6 = \square$
$4 + 5 = \square$ Think 5 + 4	$9 - 4 = \square$	$9 - 5 = \square$
$3 + 4 = \square$	$7 - 4 = \square$	$7 - 3 = \square$
$1 + 7 = \square$ Think 7 + 1	$8 - 1 = \square$	$8 - 7 = \square$
$2 + 5 = \square$	$9 - 5 = \square$	$9 - 2 = \square$
$4 + 2 = \square$	$6 - 4 = \square$	$6 - 2 = \square$
$6 + 3 = \square$	$9 - 3 = \square$	$9 - 6 = \square$
$1 + 8 = \square$	$9 - 1 = \square$	$9 - 8 = \square$
$2 + 3 = \square$	$5 - 2 = \square$	$5 - 3 = \square$

Student's Name _____　　　　　Date _____

Subtraction 1

Compute.

5	3	7	3	8
− 1	− 1	− 1	− 1	− 1

2	7	4	8	2
− 1	− 1	− 1	− 1	− 1

6	9	3	1	7
− 1	− 1	− 1	− 1	− 1

Compute.

1.　$5 - \square = 1$	2.　$3 - \square = 1$
3.　$7 - \square = 1$	4.　$7 - \square = 1$
5.　$4 - \square = 1$	6.　$5 - \square = 1$
7.　$9 - \square = 1$	8.　$3 - \square = 1$
9.　$3 - \square = 1$	10.　$7 - \square = 1$
11.　$8 - \square = 1$	12.　$6 - \square = 1$
13.　$6 - \square = 1$	14.　$8 - \square = 1$
15.　$2 - \square = 1$	16.　$1 - \square = 1$

Student's Name _____ Date _____

Subtraction 1

Compute.

You are at b2 = ☐.	<table><tr><td>**3**</td><td>5</td><td>7</td><td>4</td></tr><tr><td>**2**</td><td>3</td><td>**1**</td><td>6</td></tr><tr><td>**1**</td><td>8</td><td>2</td><td>9</td></tr><tr><td></td><td>**a**</td><td>**b**</td><td>**c**</td></tr></table>

Subtraction 1

6 7 2 9 4 3 7 3 6 2 7 2 8 3 5 9 2 3 1 7 5 8
−
_____
1 1 1 1 1 1 1 1 1 1 1 1 1 1 1 1 1 1 1 1 1 1

5 8 2 6 1 9 2 6 2 6 3 1 4 8 6 3 6 2 7 3 5 8
−
_____
1 1 1 1 1 1 1 1 1 1 1 1 1 1 1 1 1 1 1 1 1 1

6 8 3 8 3 5 3 4 8 4 9 1 5 3 6 8 7 5 3 2 4 2
−
_____
1 1 1 1 1 1 1 1 1 1 1 1 1 1 1 1 1 1 1 1 1 1

4 6 8 2 3 6 9 6 1 5 3 7 3 8 3 8 4 5 2 9 2 1
−
_____
1 1 1 1 1 1 1 1 1 1 1 1 1 1 1 1 1 1 1 1 1 1

5 8 2 8 4 7 2 1 8 3 9 1 5 2 8 3 6 3 1 7 9 3
−
_____
1 1 1 1 1 1 1 1 1 1 1 1 1 1 1 1 1 1 1 1 1 1

4 7 4 6 8 3 6 5 9 2 7 4 1 3 6 3 8 6 3 1 5 5
−
_____
1 1 1 1 1 1 1 1 1 1 1 1 1 1 1 1 1 1 1 1 1 1

168

Student's Name _____   Date _____

Counting backwards by 2

1. Count back by 2.

20	18	16			10				

21	19	17			11				

40	38				28			
28		24						

37	35									

32										

2. Fill in blanks.

9
5
6
4
8
7

$-2 =$

7

$-2 =$

5

$+1 =$

6

Student's Name _____　　Date _____

Subtracting 2

Compute.

$$4 - 2 = \square \qquad 8 - 2 = \square \qquad 5 - 2 = \square \qquad 2 - 2 = \square \qquad 6 - 2 = \square$$

$$6 - 2 = \square \qquad 3 - 2 = \square \qquad 7 - 2 = \square \qquad 8 - 2 = \square \qquad 2 - 2 = \square$$

$$3 - 2 = \square \qquad 6 - 2 = \square \qquad 2 - 2 = \square \qquad 5 - 2 = \square \qquad 8 - 2 = \square$$

Compute.

1.　5 − 2 =	2.　7 − 2 =
3.　9 − 2 =	4.　4 − 2 =
5.　4 − 2 =	6.　2 − 2 =
7.　8 − 2 =	8.　7 − 2 =
9.　2 − 2 =	10.　4 − 2 =
11.　7 − 2 =	12.　5 − 2 =
13.　3 − 2 =	14.　9 − 2 =
15.　4 − 2 =	16.　6 − 2 =

Student's Name _____   Date _____

Subtracting 2

Compute.

	You are at b2 = ☐.		3	7	9	4

	3	7	9	4
	2	3	**2**	5
	1	6	2	8
		a	b	c

Subtracting 2

Compute.

```
  6 0 2 0 7 0 8 0 5 0 2 0 4 0 3 0 6 0 8 0 2
- 2 0 2 0 2 0 2 0 2 0 2 0 2 0 2 0 2 0 2 0 2
```

```
  8 0 7 0 5 0 6 0 7 0 3 0 2 0 9 0 3 0 7 0 5
- 2 0 2 0 2 0 2 0 2 0 2 0 2 0 2 0 2 0 2 0 2
```

```
  9 0 5 0 7 0 6 0 9 0 3 0 3 0 8 0 6 0 4 0 3
- 2 0 2 0 2 0 2 0 2 0 2 0 2 0 2 0 2 0 2 0 2
```

```
  4 0 5 0 7 0 8 0 3 0 7 0 2 0 4 0 7 0 5 0 4
- 2 0 2 0 2 0 2 0 2 0 2 0 2 0 2 0 2 0 2 0 2
```

```
  7 0 3 0 8 0 9 0 4 0 3 0 4 0 5 0 7 0 2 0 3
- 2 0 2 0 2 0 2 0 2 0 2 0 2 0 2 0 2 0 2 0 2
```

```
  9 0 8 0 8 0 6 0 9 0 7 0 4 0 2 0 9 0 6 0 5
- 2 0 2 0 2 0 2 0 2 0 2 0 2 0 2 0 2 0 2 0 2
```

Student's Name _____   Date _____

Subtraction

Compute the following chess values according to the directions of knight moves.

Subtraction 2

1. Fill in blanks

| 6 |
| 9 |
| 8 |
| 7 |
| 10 |
| 12 |

- 2 →

- 2 →

- 2 →

2. Fill in each □ with an answer.

Student's Name _____   Date _____

Subtraction 2

Compute.

8	9	6	7	9
−	−	−	−	−
2	2	2	2	2

8	9	9	7	6
−	−	−	−	−
2	2	2	2	2

6	8	9	8	9
−	−	−	−	−
2	2	2	2	2

Compute.

1. $3 - \square = 2$	2. $5 - \square = 2$
3. $7 - \square = 2$	4. $7 - \square = 2$
5. $4 - \square = 2$	6. $3 - \square = 2$
7. $8 - \square = 2$	8. $8 - \square = 2$
9. $9 - \square = 2$	10. $6 - \square = 2$
11. $6 - \square = 2$	12. $4 - \square = 2$
13. $2 - \square = 2$	14. $5 - \square = 2$
15. $4 - \square = 2$	16. $9 - \square = 2$

Subtraction 2

Compute.

You are at b2 = ☐.		3	8	3	7
		2	2	**2**	6
		1	4	5	9
			a	b	c

☐ − ___ = ☐   ☐ − ___ = ☐

☐ − ___ = ☐   ☐ − ___ = ☐

☐ − ___ = ☐   ☐ − ___ = ☐

☐ − ___ = ☐   ☐ − ___ = ☐

☐ − ___ = ☐   ☐ − ___ = ☐

☐ − ___ = ☐   ☐ − ___ = ☐

☐ − ___ = ☐   ☐ − ___ = ☐

☐ − ___ = ☐   ☐ − ___ = ☐

Subtracting 2
Counting backwards

1. Count back by 3.

30	27	24			15				

31	28								4

40	37								

50		41							

36			24						

42									

2. Fill in blanks.

6
4
9
5
7
8

$- 3 =$

$+ 2 =$

$- 1 =$

Subtraction 3

1. Compute.

$$
\begin{array}{ccccc}
7 & 5 & 8 & 3 & 8 \\
-\ 3 & -\ 3 & -\ 3 & -\ 3 & -\ 3 \\
\square & \square & \square & \square & \square
\end{array}
$$

$$
\begin{array}{ccccc}
5 & 7 & 9 & 3 & 4 \\
-\ 3 & -\ 3 & -\ 3 & -\ 3 & -\ 3 \\
\square & \square & \square & \square & \square
\end{array}
$$

$$
\begin{array}{ccccc}
6 & 8 & 4 & 9 & 8 \\
-\ 3 & -\ 3 & -\ 3 & -\ 3 & -\ 3 \\
\square & \square & \square & \square & \square
\end{array}
$$

2. Compute.

1. $8 - 3 =$	2. $5 - 3 =$
3. $7 - 3 =$	4. $8 - 3 =$
5. $3 - 3 =$	6. $3 - 3 =$
7. $9 - 3 =$	8. $9 - 3 =$
9. $5 - 3 =$	10. $5 - 3 =$
11. $7 - 3 =$	12. $4 - 3 =$
13. $4 - 3 =$	14. $7 - 3 =$
15. $6 - 3 =$	16. $7 - 3 =$

Subtraction 3

```
6 8 4 5 6 4 4 7 6 9 7 9 6 5 8 6 9 6 5 9 6 9
- 3 3 3 3 3 3 3 3 3 3 3 3 3 3 3 3 3 3 3 3 3 3
```

```
6 4 8 7 9 7 9 5 5 4 4 3 8 6 4 8 5 8 5 3 8 5
- 3 3 3 3 3 3 3 3 3 3 3 3 3 3 3 3 3 3 3 3 3 3
```

```
7 4 3 8 6 9 5 7 5 4 7 5 8 4 8 6 4 9 5 9 6 7
- 3 3 3 3 3 3 3 3 3 3 3 3 3 3 3 3 3 3 3 3 3 3
```

```
5 3 7 4 3 8 6 9 5 8 5 8 5 4 9 6 7 9 6 9 6 4
- 3 3 3 3 3 3 3 3 3 3 3 3 3 3 3 3 3 3 3 3 3 3
```

```
7 5 8 9 3 7 9 7 9 6 8 5 3 7 4 7 6 4 7 5 3 7
- 3 3 3 3 3 3 3 3 3 3 3 3 3 3 3 3 3 3 3 3 3 3
```

```
4 2 6 3 7 3 8 6 3 7 5 8 6 9 5 9 5 6 4 8 4 8
- 3 3 3 3 3 3 3 3 3 3 3 3 3 3 3 3 3 3 3 3 3 3
```

Subtraction 3

Compute.

You are at b2 ☐.	

3	6	3	7
2	5	**3**	6
1	9	4	8
	a	b	c

Student's Name _____

Date _____

Subtraction 3

1. Compute.

6	4	7	6	9
− __	− __	− __	− __	− __
3	3	3	3	3

3	7	9	8	4
− __	− __	− __	− __	− __
3	3	3	3	3

5	7	4	9	6
− __	− __	− __	− __	− __
3	3	3	3	3

2. Compute.

1. $5 - \square = 3$		2. $7 - \square = 3$	
3. $8 - \square = 3$		4. $3 - \square = 3$	
5. $4 - \square = 3$		6. $7 - \square = 3$	
7. $5 - \square = 3$		8. $8 - \square = 3$	
9. $9 - \square = 3$		10. $5 - \square = 3$	
11. $5 - \square = 3$		12. $6 - \square = 3$	
13. $7 - \square = 3$		14. $9 - \square = 3$	
15. $3 - \square = 3$		16. $4 - \square = 3$	

Subtraction 3

Compute.

3	4	8	3
2	7	3	6
1	9	5	8
	a	b	c

You are at b2 □.

− ___ = □

− ___ = □

− ___ = □

− ___ = □

− ___ = □

− ___ = □

− ___ = □

− ___ = □

− ___ = □

− ___ = □

− ___ = □

− ___ = □

− ___ = □

− ___ = □

− ___ = □

− ___ = □

Student's Name _____   Date _____

Subtraction 3

Compute.

```
  6 4 7 4 7 9 4 3 8 4 7 9 6 4 8 6 5 4 6 8 9 5
- 3 1 4 1 4 6 1 0 5 1 4 6 3 1 5 3 2 1 3 5 6 2
```

```
  5 4 7 5 3 7 8 6 9 7 5 8 6 5 3 8 7 5 9 6 4 9
- 2 1 4 2 0 4 5 3 6 4 2 5 3 2 0 5 4 2 6 3 1 6
```

```
  6 8 7 5 9 6 9 6 4 5 8 6 9 7 5 6 4 7 5 5 8 6
- 3 5 4 2 6 3 6 3 1 2 5 3 6 4 2 3 1 4 2 2 5 3
```

```
  6 4 7 4 3 7 8 9 5 4 3 9 6 5 4 9 6 7 4 7 9 7
- 3 1 4 1 0 4 5 6 2 1 0 6 3 2 1 6 3 4 1 4 6 4
```

```
  6 4 6 7 8 5 3 7 6 5 9 7 6 9 4 6 7 9 4 3 7 8
- 3 1 3 4 5 2 0 4 3 2 6 4 3 6 1 3 4 6 1 0 4 5
```

```
  6 4 3 8 6 3 5 9 5 6 3 6 7 5 8 4 9 5 7 5 6 8
- 3 1 0 5 3 0 2 6 2 3 0 3 4 2 5 1 6 2 4 2 3 5
```

Student's Name _____　　　　　Date _____

Calculate with borrowing.

1. Regrouping

$11 - 9$ $= 10 + 1\square - 9$ $= (10 - 9) + 1$ (use 10 to subtract 9 first) $= 1 + 1$ $= \square\, 2$	Step 1: $10 - 9 = 1$ Step 2: $1 + 1 = 2$ $1\,1 = 10 + 1$ $\underline{-\ 9}$ $\square\, 2$	
$11 - 8$ $= 10 + \square\, 1 - 8$ $= (10 - 8) + 1$ $= 2 + 1$ $= \square\, 3$	Step 1: $10 - 8 = 2$ Step 2: $2 + 1 = 3$ $1\,1 = 10 + 1$ $\underline{-\ 8}$ $\square\, 3$	

2. Flip the left diagram to the right-hand side using the symmetry line.

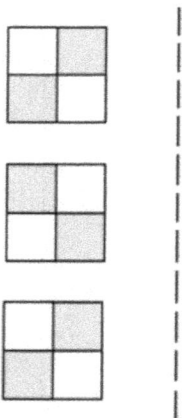

1. Calculate.

1 5	1 4	1 8	1 3	1 7
− 9	− 9	− 9	− 9	− 9
□	□	□	□	□

1 0	1 6	1 2	1 7	1 5
− 9	− 9	− 9	− 9	− 9
□	□	□	□	□

1 8	1 1	1 9	1 3	1 7
− 9	− 9	− 9	− 9	− 9
□	□	□	□	□

2. Compute.

1.	$16 - 9 =$	2.	$10 - 9 =$
3.	$18 - 9 =$	4.	$18 - 9 =$
5.	$16 - 9 =$	6.	$11 - 9 =$
7.	$19 - 9 =$	8.	$17 - 9 =$
9.	$10 - 9 =$	10.	$12 - 9 =$
11.	$13 - 9 =$	12.	$13 - 9 =$
13.	$17 - 9 =$	14.	$18 - 9 =$
15.	$15 - 9 =$	16.	$19 - 9 =$

Compute.

You are at b2 = ☐.	<table><tr><td>**3**</td><td>11</td><td>17</td><td>15</td></tr><tr><td>**2**</td><td>16</td><td>**9**</td><td>12</td></tr><tr><td>**1**</td><td>13</td><td>18</td><td>14</td></tr><tr><td></td><td>**a**</td><td>**b**</td><td>**c**</td></tr></table>

– ☐ = 17-9=8	– ☐ = 12-9=3
– ☐ = 18-9=9	– ☐ =14-9=5
– ☐ = 12-9=3	– ☐ =15-9=6
– ☐ =16-9=7	– ☐ =13-9=4
– ☐ =15-9=6	– ☐ =17-9=8
– ☐ =14-9=5	– ☐ =11-9=2
– ☐ =13-9=4	– ☐ =18-9=9
– ☐ =11-9=2	– ☐ =16-9=7

Compute.

1 3 1 4 1 5 1 7 1 3 1 1 1 6 1 2 1 7 1 3 1 2
− 0 9 0 9 0 9 0 9 0 9 0 9 0 9 0 9 0 9 0 9 0 9

1 3 1 0 1 5 1 3 1 6 1 2 1 6 1 4 1 7 1 1 1 8
− 0 9 0 9 0 9 0 9 0 9 0 9 0 9 0 9 0 9 0 9 0 9

1 3 1 5 1 2 1 6 1 3 1 8 1 5 1 2 1 1 1 7 1 2
− 0 9 0 9 0 9 0 9 0 9 0 9 0 9 0 9 0 9 0 9 0 9

1 4 1 3 1 6 1 3 1 8 1 4 1 5 1 9 1 2 1 0 1 1
− 0 9 0 9 0 9 0 9 0 9 0 9 0 9 0 9 0 9 0 9 0 9

1 4 1 1 1 2 1 0 1 5 1 2 1 5 1 7 1 4 1 7 1 2
− 0 9 0 9 0 9 0 9 0 9 0 9 0 9 0 9 0 9 0 9 0 9

1 2 1 1 1 0 1 3 1 6 1 5 1 8 1 3 1 4 1 5 1 8
− 0 9 0 9 0 9 0 9 0 9 0 9 0 9 0 9 0 9 0 9 0 9

1. Calculate.

$$12 - 8 = \square \qquad 15 - 8 = \square \qquad 16 - 8 = \square \qquad 13 - 8 = \square \qquad 17 - 8 = \square$$

$$14 - 8 = \square \qquad 15 - 8 = \square \qquad 17 - 8 = \square \qquad 14 - 8 = \square \qquad 16 - 8 = \square$$

$$11 - 8 = \square \qquad 14 - 8 = \square \qquad 10 - 8 = \square \qquad 16 - 8 = \square \qquad 13 - 8 = \square$$

2. Compute.

1.　$10 - 8 =$	2.　$16 - 8 =$
3.　$15 - 8 =$	4.　$13 - 8 =$
5.　$18 - 8 =$	6.　$18 - 8 =$
7.　$17 - 8 =$	8.　$12 - 8 =$
9.　$15 - 8 =$	10.　$11 - 8 =$
11.　$13 - 8 =$	12.　$15 - 8 =$
13.　$14 - 8 =$	14.　$17 - 8 =$
15.　$11 - 8 =$	16.　$10 - 8 =$

Student's Name _____　　　　Date _____

Compute.

You are at b2 = □.

3	13	11	14
2	15	**8**	12
1	16	13	17
	a	**b**	**c**

□ − □ = □ − □ =

□ − □ = □ − □ =

□ − □ = □ − □ =

□ − □ = □ − □ =

□ − □ = □ − □ =

□ − □ = □ − □ =

□ − □ = □ − □ =

□ − □ = □ − □ =

189

Compute.

$$1\ 0\ 1\ 3\ 1\ 2\ 1\ 5\ 1\ 3\ 1\ 5\ 1\ 1\ 1\ 4\ 1\ 4\ 1\ 7\ 1\ 6$$
$$-\ 0\ 8\ 0\ 8\ 0\ 8\ 0\ 8\ 0\ 8\ 0\ 8\ 0\ 8\ 0\ 8\ 0\ 8\ 0\ 8\ 0\ 8$$

$$1\ 2\ 1\ 0\ 1\ 4\ 1\ 7\ 1\ 3\ 1\ 7\ 1\ 2\ 1\ 7\ 1\ 8\ 1\ 3\ 1\ 0$$
$$-\ 0\ 8\ 0\ 8\ 0\ 8\ 0\ 8\ 0\ 8\ 0\ 8\ 0\ 8\ 0\ 8\ 0\ 8\ 0\ 8\ 0\ 8$$

$$1\ 2\ 1\ 4\ 1\ 7\ 1\ 3\ 1\ 8\ 1\ 4\ 1\ 1\ 1\ 6\ 1\ 2\ 1\ 5\ 1\ 2$$
$$-\ 0\ 8\ 0\ 8\ 0\ 8\ 0\ 8\ 0\ 8\ 0\ 8\ 0\ 8\ 0\ 8\ 0\ 8\ 0\ 8\ 0\ 8$$

$$1\ 2\ 1\ 4\ 1\ 7\ 1\ 8\ 1\ 1\ 1\ 4\ 1\ 0\ 1\ 2\ 1\ 5\ 1\ 3\ 1\ 8$$
$$-\ 0\ 8\ 0\ 8\ 0\ 8\ 0\ 8\ 0\ 8\ 0\ 8\ 0\ 8\ 0\ 8\ 0\ 8\ 0\ 8\ 0\ 8$$

$$1\ 2\ 1\ 5\ 1\ 3\ 1\ 4\ 1\ 6\ 1\ 0\ 1\ 6\ 1\ 3\ 1\ 4\ 1\ 7\ 1\ 2$$
$$-\ 0\ 8\ 0\ 8\ 0\ 8\ 0\ 8\ 0\ 8\ 0\ 8\ 0\ 8\ 0\ 8\ 0\ 8\ 0\ 8\ 0\ 8$$

$$1\ 0\ 1\ 2\ 1\ 7\ 1\ 5\ 1\ 7\ 1\ 3\ 1\ 4\ 1\ 2\ 1\ 5\ 1\ 6\ 1\ 2$$
$$-\ 0\ 8\ 0\ 8\ 0\ 8\ 0\ 8\ 0\ 8\ 0\ 8\ 0\ 8\ 0\ 8\ 0\ 8\ 0\ 8\ 0\ 8$$

1. Count back by 10.

100	90	80						

92	82				32			

97		67						

91						21		

95		75						

98			58					

2. Fill in blanks.

36
25
47
51
39
43

$- 10 =$

$+ 2 =$

$- 10 =$

1. Calculate.

1 4	1 6	1 7	1 2	1 8
− 1 0	− 1 0	− 1 0	− 1 0	− 1 0

1 3	1 3	1 9	1 4	1 2
− 1 0	− 1 0	− 1 0	− 1 0	− 1 0

1 8	1 7	1 8	1 2	1 1
− 1 0	− 1 0	− 1 0	− 1 0	− 1 0

2. Compute.

1.	$16 - \square = 10$	2.	$12 - \square = 10$
3.	$14 - \square = 10$	4.	$18 - \square = 10$
5.	$17 - \square = 10$	6.	$19 - \square = 10$
7.	$13 - \square = 10$	8.	$14 - \square = 10$
9.	$10 - \square = 10$	10.	$16 - \square = 10$
11.	$11 - \square = 10$	12.	$13 - \square = 10$
13.	$18 - \square = 10$	14.	$15 - \square = 10$
15.	$14 - \square = 10$	16.	$17 - \square = 10$

Compute.

You are at b2 = ☐ .	**3** 12 18 11 **2** 16 **10** 14 **1** 15 17 19 **a** **b** **c**

− ___ = ☐   − ___ = ☐

− ___ = ☐   − ___ = ☐

− ___ = ☐   − ___ = ☐

− ___ = ☐   − ___ = ☐

− ___ = ☐   − ___ = ☐

− ___ = ☐   − ___ = ☐

− ___ = ☐   − ___ = ☐

− ___ = ☐   − ___ = ☐

Compute.

1 3	1 4	1 5	1 7	1 3	1 1	1 6	1 2	1 7	1 3	1 2
− 0	0	0	0	0	0	0	0	0	0	0 2
1 0 1 0 1 0 1 0 1 0 1 0 1 0 1 0 1 0 1 0 1 0

1 3	1 0	1 5	1 3	1 6	1 2	1 6	1 4	1 7	1 1	1 8
− 0	0	0	0	0	0	0	0	0	0	0 8
1 0 1 0 1 0 1 0 1 0 1 0 1 0 1 0 1 0 1 0 1 0

1 3	1 5	1 2	1 6	1 3	1 8	1 5	1 2	1 1	1 7	1 2
− 0	0	0	0	0	0	0	0	0	0	0 2
1 0 1 0 1 0 1 0 1 0 1 0 1 0 1 0 1 0 1 0 1 0

1 4	1 3	1 6	1 3	1 8	1 4	1 5	1 9	1 2	1 0	1 1
− 0	0	0	0	0	0	0	0	0	0	0 1
1 0 1 0 1 0 1 0 1 0 1 0 1 0 1 0 1 0 1 0 1 0

1 4	1 1	1 2	1 0	1 5	1 2	1 5	1 7	1 4	1 7	1 2
− 0	0	0	0	0	0	0	0	0	0	0 2
1 0 1 0 1 0 1 0 1 0 1 0 1 0 1 0 1 0 1 0 1 0

1 2	1 1	1 0	1 3	1 6	1 5	1 8	1 3	1 4	1 5	1 8
− 0	0	0	0	0	0	0	0	0	0	0 8
1 0 1 0 1 0 1 0 1 0 1 0 1 0 1 0 1 0 1 0 1 0

1. Complete the table.

14 minus 4 is ___	14 minus 10 is ___	___ – 4 = 10	___ – 10 = 4
15 minus 6 is ___	15 minus 6 is ___	15 – ___ = 6	15 – 9 = ___
15 minus 8 is ___	15 minus 7 is ___	15 – 7 = ___	15 – ___ = 8
16 minus 7 is ___	16 minus 9 is ___	___ – 9 = 7	16 – ___ = 9
16 minus 6 is ___	16 minus 10 is ___	___ – 6 = 10	16 – ___ = 6
13 minus 9 is ___	13 minus 4 is ___	___ – 9 = 4	13 – ___ = 9
14 minus 8 is ___	14 minus 6 is ___	___ – 8 = 6	14 – ___ = 8
18 minus 9 is ___	18 minus 9 is ___	___ – 9 = 9	18 – ___ = 9

2. Replace each ? with a number.

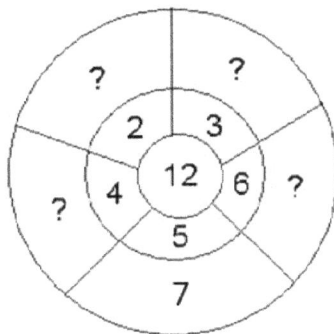

Student's Name _____ Date _____

1. Calculate.

$11 - 2 = \square$	$17 - 7 = \square$
$15 - 2 = \square$	$18 - 9 = \square$
$17 - 3 = \square$	$12 - 9 = \square$
$12 - 5 = \square$	$17 - 7 = \square$
$13 - 4 = \square$	$13 - 9 = \square$
$13 - 6 = \square$	$14 - 8 = \square$
$14 - 5 = \square$	$17 - 6 = \square$
$16 - 5 = \square$	$14 - 5 = \square$
$15 - 6 = \square$	$15 - 7 = \square$
$19 - 6 = \square$	$15 - 8 = \square$
$18 - 7 = \square$	$16 - 9 = \square$

2. Can you find the answers to the question marks in the following diagram?

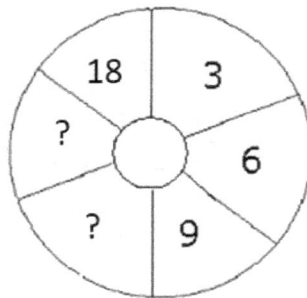

Student's Name _____　　　Date _____

1. Calculate.

$$
\begin{array}{ccccc}
13 & 17 & 12 & 18 & 15 \\
-9 & -9 & -9 & -9 & -9
\end{array}
$$

$$
\begin{array}{ccccc}
14 & 12 & 17 & 15 & 16 \\
-9 & -9 & -9 & -9 & -9
\end{array}
$$

$$
\begin{array}{ccccc}
13 & 17 & 11 & 10 & 14 \\
-9 & -9 & -9 & -9 & -9
\end{array}
$$

2. Compute.

1.　　$12 - \square = 9$	2.　　$11 - \square = 9$
3.　　$17 - \square = 9$	4.　　$17 - \square = 9$
5.　　$14 - \square = 9$	6.　　$14 - \square = 9$
7.　　$16 - \square = 9$	8.　　$18 - \square = 9$
9.　　$10 - \square = 9$	10.　$12 - \square = 9$
11.　$14 - \square = 9$	12.　$16 - \square = 9$
13.　$17 - \square = 9$	14.　$11 - \square = 9$
15.　$18 - \square = 9$	16.　$15 - \square = 9$

Student's Name _____　　　Date _____

Compute.

You are at b2 = ☐ .	

3	19	11	16
2	14	**9**	12
1	17	18	15
	a	**b**	**c**

☐ − ___ = ☐

☐ − ___ = ☐

☐ − ___ = ☐

☐ − ___ = ☐

☐ − ___ = ☐

☐ − ___ = ☐

☐ − ___ = ☐

☐ − ___ = ☐

☐ − ___ = ☐

☐ − ___ = ☐

☐ − ___ = ☐

☐ − ___ = ☐

☐ − ___ = ☐

☐ − ___ = ☐

Student's Name _____   Date _____

1. Calculate.

1 1	1 2	1 3	1 4	1 5
− 2	− 3	− 4	− 5	− 6
□	□	□	□	□

1 6	1 7	1 8	1 2	1 2
− 7	− 8	− 9	− 3	− 4
□	□	□	□	□

1 1	1 4	1 6	1 7	1 8
− 5	− 6	− 7	− 8	− 9
□	□	□	□	□

1 3	1 2	1 4	1 5	1 6
− 4	− 5	− 6	− 7	− 8
□	□	□	□	□

2. Fill in the blank with a number.

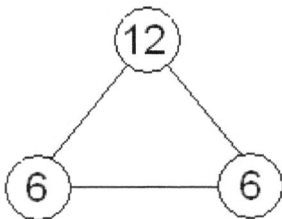

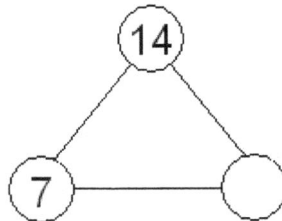

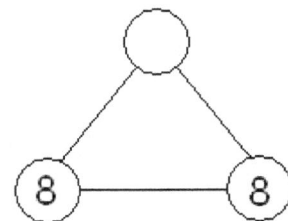

Ho Math Chess 何数棋谜　低年级棋谜式数学

Student's Name _____ Date _____

Compute.

```
 1 5 1 2 1 7 1 2 1 7 1 4 1 6 1 8 1 3 1 2 1 0
-0   0   0   0   0   0   0   0   0   0   0
 0 9 0 9 0 9 0 9 0 9 0 9 0 9 0 9 0 9 0 9 0 9
```

```
 1 2 1 6 1 4 1 2 1 6 1 7 1 4 1 2 1 8 1 2 1 3
-0   0   0   0   0   0   0   0   0   0   0
 0 9 0 9 0 9 0 9 0 9 0 9 0 9 0 9 0 9 0 9 0 9
```

```
 1 5 1 5 1 2 1 6 1 3 1 6 1 3 1 0 1 4 1 7 1 3
-0   0   0   0   0   0   0   0   0   0   0
 0 9 0 9 0 9 0 9 0 9 0 9 0 9 0 9 0 9 0 9 0 9
```

```
 1 2 1 2 1 6 1 3 1 5 1 7 1 6 1 1 1 0 1 0 1 4
-0   0   0   0   0   0   0   0   0   0   0
 0 9 0 9 0 9 0 9 0 9 0 9 0 9 0 9 0 9 0 9 0 9
```

```
 1 6 1 5 1 8 1 2 1 4 1 3 1 8 1 3 1 4 1 5 1 7
-0   0   0   0   0   0   0   0   0   0   0
 0 9 0 9 0 9 0 9 0 9 0 9 0 9 0 9 0 9 0 9 0 9
```

```
 1 2 1 6 1 7 1 5 1 3 1 4 1 7 1 5 1 8 1 6 1 2
-0   0   0   0   0   0   0   0   0   0   0
 0 9 0 9 0 9 0 9 0 9 0 9 0 9 0 9 0 9 0 9 0 9
```

Student's Name _____　　Date _____

Mixed computation

✳ ♚ = 0, ♙ ♟ = 1, ✠ ♝ = 3, ✛ ♞ = 3,　✛ ♜ = 5, ✳ ♛ = 9

Spatial relation and computing

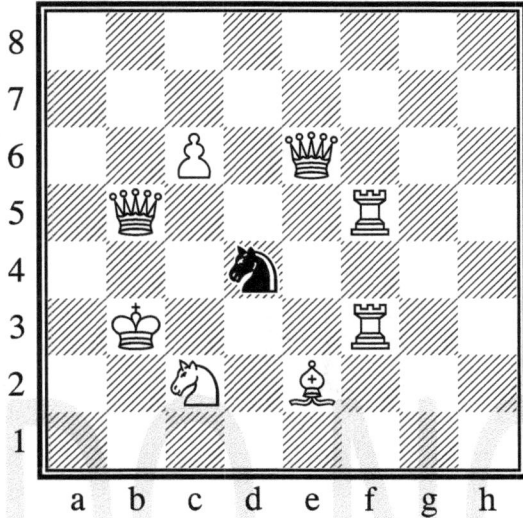

Compute the following chess values according to the directions of knight moves.

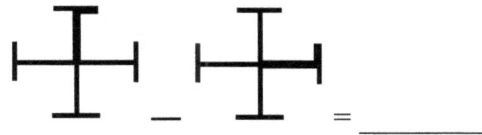

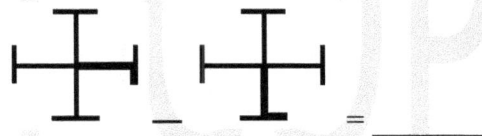

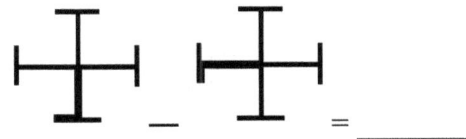

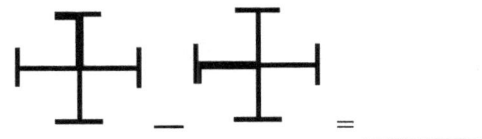

⊢ − ⊤ = _____

⊢ − ⊥ = _____

⊢ − ⊢ = _____

⊢ − ⊢ = _____

Student's Name _____   Date _____

Spatial relation and computing

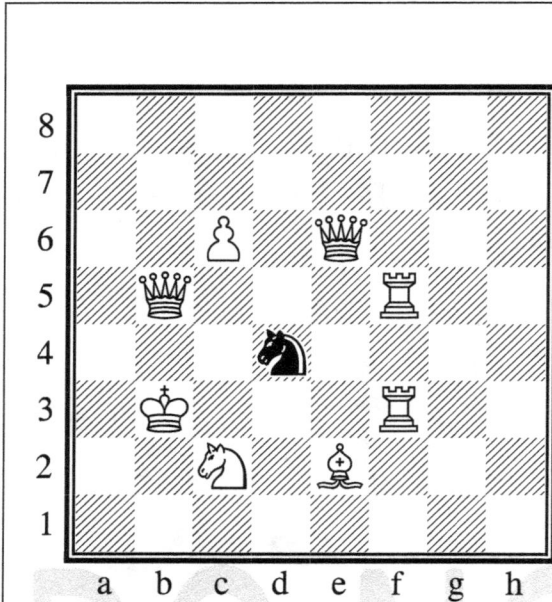

Compute the following chess values according to the directions of knight moves.

$$\boxed{ } + \boxed{ } = \underline{\qquad}$$

$$\boxed{ } + \boxed{ } = \underline{\qquad}$$

$$\boxed{ } + \boxed{ } = \underline{\qquad}$$

$$\boxed{ } + \boxed{ } = \underline{\qquad}$$

Spatial relation and boundary

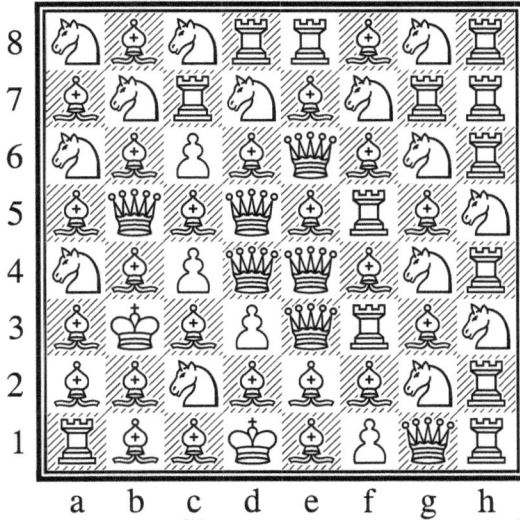

Find the number of chess pieces on the path of Qe4.

	♚	♞	♙	♛
# of chess pieces				
	♖	♗	♔	♘
# of chess pieces				

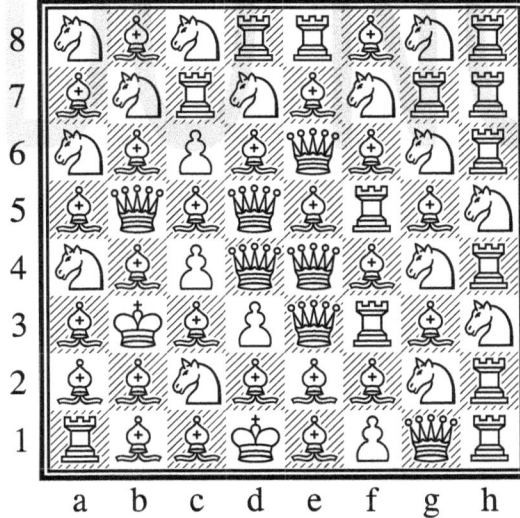

Find the number of chess pieces on the path of non-Qe4.

	♔	♞	♙	♛
# of chess pieces				
	♖	♗	♔	♘
# of chess pieces				

Spatial relation and Computing

	Find Answers
6 4 6 3 6 3 8 8 6 0 3 4 ♛ 2 3 3 4 6 3 2 6 7 9 7 2	Circle the two numbers next to each other in moves as indicated by the chess piece on the left chess diagram such that the two numbers make the difference of 1. Find as many answers as you can
6 4 6 3 6 3 1 7 3 2 4 9 ♞ 6 7 3 8 8 6 0 3 2 5 1 4	Circle the two numbers next to each other in moves as indicated by the chess piece on the left chess diagram such that the two numbers make the difference of 1. Find as many answers as you can

Student's Name _____　Date _____

Spatial relation and Computing

6	3	9	6	10
3	8	7	8	5
4	6	♛	5	7
6	8	3	7	9
2	10	1	6	8

Circle the two numbers next to each other in moves as indicated by the chess piece on the left chess diagram such that the two numbers make the difference of 2. Find as many answers you can

6	3	9	6	10
3	8	6	8	5
4	6	♜	5	8
6	8	3	7	9
2	10	0	6	8

Circle the two numbers next to each other in moves as indicated by the chess piece on the left chess diagram such that the two numbers make the difference of 3. Find as many answers you can

Spatial relation and Computing

	Find Answers
  6 4 6 3 3 3 8 8 3 0 3 4 ♛ 2 5 3 4 6 3 2 6 7 1 7 2 	Circle the two numbers next to each other in moves as indicated by the chess piece on the left chess diagram such that the two numbers make the sum of 7. Find as many answers you can
  6 4 6 3 6 4 1 7 3 5 4 9 ♞ 6 7 3 8 8 9 6 3 2 5 1 4 	Circle the two numbers next to each other in moves as indicated by the chess piece on the left chess diagram such that the two numbers make the sum of 8 Find as many answers you can

Spatial relation and Computing

1	3	9	6	2
3	8	0	8	5
3	6	♕	5	4
6	8	6	7	9
1	10	3	6	3

Circle the two numbers next to each other in moves as indicated by the chess piece on the left chess diagram that such that the two numbers make the sum of 9. Find as many answers you can

6	3	1	6	10
3	8	6	8	5
4	2	♖	5	1
6	8	3	7	9
2	10	2	6	8

Circle the two numbers next to each other in moves as indicated by the chess piece on the left chess diagram such that the two numbers make the sum of 6. Find as many answers you can

Spatial relation and Computing

1	3	9	6	2
3	8	0	8	5
3	6	6	5	4
6	8	6	7	9
1	10	3	6	3

Circle the two numbers next to each other in queen's moves that make the sum of 9. Find as many answers you can.

6	3	1	6	10
3	8	6	8	5
4	2	4	5	1
6	8	3	7	9
2	10	2	6	8

Circle the two numbers next to each other in queen's moves that make the sum of 10. Find as many answers you can.

Student's Name _____ Date _____

Spatial relation and Computing

1	3	9	6	2
3	8	0	8	5
3	6	♞	5	4
6	8	6	7	9
1	10	3	6	3

Add two diagonal numbers next to each other in knight moves and list the results from the smallest to the largest.

6	3	1	6	10
3	8	6	8	5
4	2	♞	5	1
6	8	3	7	9
2	10	2	6	8

Add two diagonal numbers next to each other in knight moves and list the results from the smallest to the largest.

Add up the points ♗ at b2 has travelled safely to reach ✖? _____

Add up the points ♞ at b8 has travelled safely to reach b2? _____

Student's Name _____　　　　　Date _____

Relation and counting

※ ♔ = 0, ♕ ♙ = 1, ✖ ♗ = 3, ✚ ♘ = 3, ✚ ♖ = 5, ※ ♕ = 9

Table	Direction	Computation
♙ 7 / ♘ 4 / ♟ 6 / ※ 8 0	(arrows)	___ + ___ = ___ ___ − ___ = ___
♟ 3 4 / 7 ♖ 6 / ♛ 8 ♚	(arrows)	___ + ___ = ___ ___ − ___ = ___
1 ♗ 4 / 7 6 4 / ♛ 8 ♞	(arrows)	___ + ___ = ___ ___ − ___ = ___
1 ✳ 4 / 7 9 6 / 6 ♗ ♞	(arrows)	___ − ___ = ___ ___ + ___ = ___

Student's Name _____　　Date _____

Relation and counting

❋ ♔ = 0, ♕ ♙ = 1, ✠ ♗ = 3, ✤ ♘ = 3, ✛ ♖ = 5, ❋ ♛ = 9

Table			Direction	Computation
♙	♞	♞		____ + ____ = ____
7	4	6		
❋	8	0		____ − ____ = ____
♟	2	4		____ + ____ = ____
7	♖	6		
♛	8	♚		____ − ____ = ____
1	♗	4		____ + ____ = ____
7	6	3		
♛	8	♞		____ − ____ = ____
1	❋	4		____ + ____ = ____
7	7	6		
6	♗	♞		____ − ____ = ____

Ho Math, Chess, and Puzzles for Grade 1 and Under

Ho Math Chess 何数棋谜 低年级棋谜式数学

Student's Name _____   Date _____

Relation and counting

♔ = 0, ♙ = 1, ♗ = 3, ♘ = 3, ♖ = 5, ♕ = 9

♚	♟	2
6	1	3
4	7	♕

A. _____ + _____ = _____

B. _____ − _____ = _____

0	1	2
♕	7	♞
4	8	4

C. _____ − _____ = _____

D. _____ + _____ = _____

♚	1	2	3	♞
6	♙	♗	4	3
4	7	♕	6	5
♞	♖	8	0	3
2	3	6	1	♟

E. _____ − _____ = _____

F. _____ + _____ = _____

3	♙	2	♗	♗
6	6	♞	4	♞
4	7	♕	6	♖
♞	♖	8	0	3
2	♗	6	2	♙

G. _____ − _____ = _____

H. _____ + _____ = _____

A	B	C
♜	D	E
F	G	H

I. _____ − _____ = _____

J. _____ + _____ = _____

Student's Name _____   Date _____

Math and Chess Puzzle

You are at c3.

	a	b	c	d	e
5		••/••	✳	•••/••	
4	•••(dots)	3	4	6	••/••
3	✳	6	10 13 / 15 17	8	✳
2	•	7	8	9	•••
1		•	✳	•••	

The points of chess pieces are as follows:

✳ = 0

↓ = 1

⤢ = 3

┼ = 3

↔↕ = 5

✳(boxed) = 9

Chess notation and computation

✳ ♔ = 0, ♕ ♙ = 1, ✠ ♝ = 3, ♞ = 3, ♟ ♖ = 5, ✳ ♛ = 9

	a	b	c	d
8				
7				
6				
5				
4				
3				
2				
1				

a1 – b1 = ____	c1 – c6 = ____	d4 – d7 = ____
a5 – c1 = ____	b1 + b4 = ____	d4 + c5 = ____
b5 – b4 = ____	a7 – c6 = ____	a6 – d7 = ____
a8 – c7 = ____	c5 – b6 = ____	c4 + d7 = ____
a3 – c1 = ____	c1 – c7 = ____	a5 – c7 = ____
a1 + b5 = ____	c4 + c6 = ____	d4 + d7 = ____
a4 – c8 = ____	c5 – d6 = ____	c4 + d7 = ____

Student's Name _____   Date _____

Addition

Chess diagram	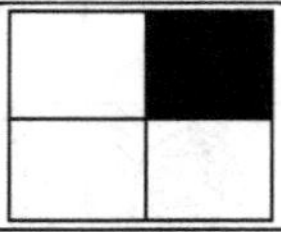	
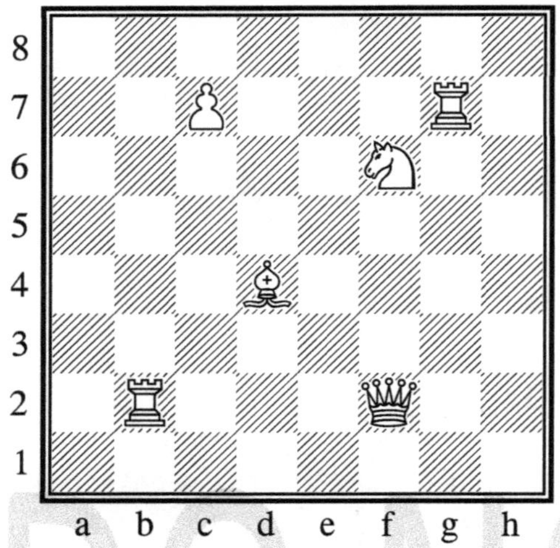	What are the total points of White?	What are the total points of White?

Fill in each ☐ by a number.

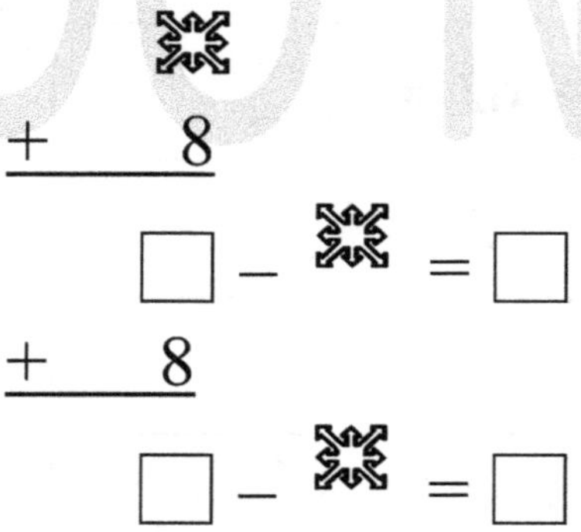

    �֎

+   8
_____

☐ − �֎ = ☐

+   8
_____

☐ − ✖ = ☐

The left is a computation puzzle created by SCL.

216

Student's Name _____ Date _____

Addition

?		8		?
	8	7	6	
4	11	15	12	3
	10	5	9	
5		?		?

Add two numbers in moves of ✳ but converging to 15. Replace each ? with a number.

9		?		?
	8	7	6	
4	?	17	12	?
	10	5	9	
?		?		?

Add two numbers in moves of ✳ but converging to 17. Replace each ? with a number.

Subtraction

Chess diagram	How many points are White more than Black? Answer: _____	How many points are White more than Black? Answer: _____
Fill in each ☐ by a number.		

Student's Name _____   Date _____

Largest and smallest

Replace each ? with a number.

In all moves of ✺ at the centre of 13 (6 + ? = 13), add the largest number and the smallest number found. ._____.

?		?		?
	8	9	6	
?	10	13	6	?
	?	5	9	
1		?		?

Replace each ? with a number.

In all moves of ✺ at the centre of 16 (6 + ? = 16), add the largest number and the smallest number found. ._____.

9		?		?
	8	7	6	
4	?	16	3	?
	10	5	9	
?		?		?

Largest and smallest

?		?		?
	8	9	6	
?	10	19	6	?
	?	5	9	
1		?		?

Replace each ? with a number.

In all moves of ✴ at the centre of 19 (6 + ? = 19), find the difference between the largest number and the smallest number found. ._____.

8		?		?
	13	7	6	
?	17	21	3	?
	10	5	9	
?		?		?

Replace each ? with a number.

In all moves of ✴ at the centre of 21 (6 + ? = 21), find the difference between the largest number and the smallest number found. ._____.

Student's Name _____   Date _____

Magic Knight

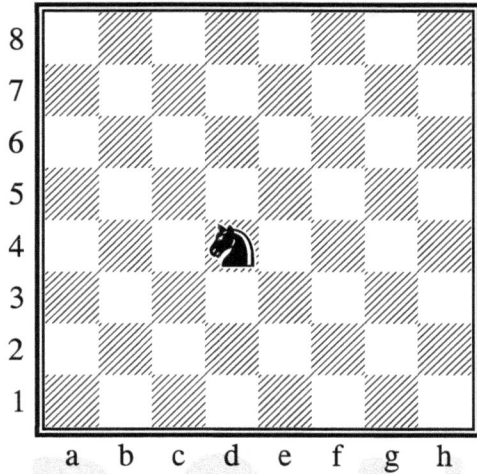

Chess Diagram	
	Place numbers from 1 to 9 starting at Nd4 and along its all possible squares of one-move. The sum of its L shaped squares (3 squares) must all be equal to 15.
	Replace each ? by a number such that any 2 numbers multiplied together is 24 following the following directions .

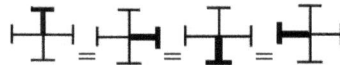

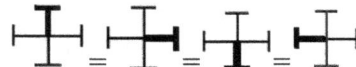

Set

Chess Puzzle Samples	
Cross mark (X) the squares where both chess pieces can move to (intersect).	The squares of ✳ intersecting with ⊤ are _____.
Cross mark (X) the squares where both chess pieces can move to (intersect).	The squares of ✳ intersecting with ⊤ are _____.

Set

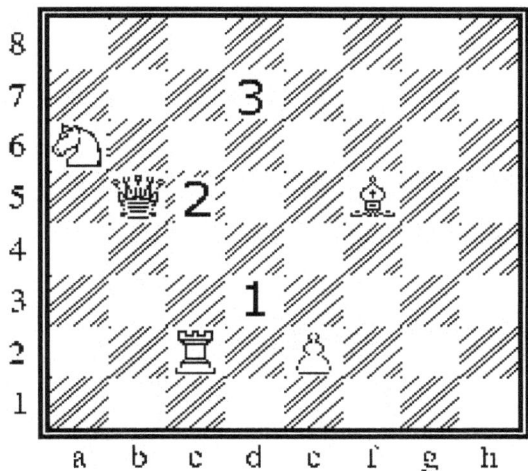

When compared to the above diagram, what is the total the value of all chess pieces, which occupy all ✕'s or ✕'s in the diagram below?

Answer: _____

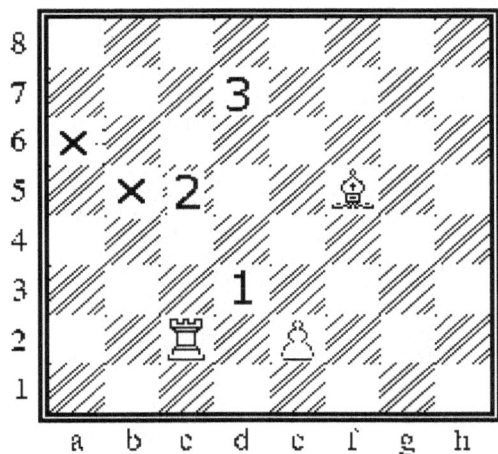

Use the left top diagram, do the following.

Fill in each ☐ with an answer.

♖ ∩ ♘ + ♕ ∩ ♖ = ☐

♞ ∩ ♛ + ♛ ∩ ♟ = ☐

♕ ∩ ♗ + ♗ ∩ ♟ = ☐

c2 + e2 = ☐ + ☐ = ☐

f5 + b5 = ☐ + ☐ = ☐

d7 + a6 = ☐ + ☐ = ☐

d3 + c5 + f5

= ☐ + ☐ + ☐ = ☐

d7 + b5 + a6

= ☐ + ☐ + ☐ = ☐

Student's Name _____　　　　Date _____

When is chess distance not real distance?

Learning distance in an abstract way?

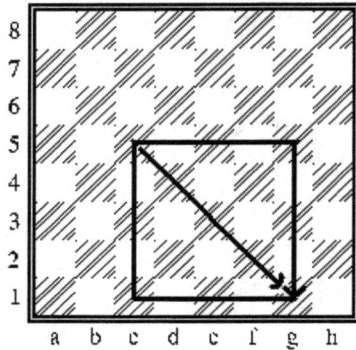

The distance from c5 to g1 takes the same number of squares to reach from g5 to g1. How many squares are there? _____

In fact, the real distance from c5 to g1 is longer than g5 to g1.

Knight jump

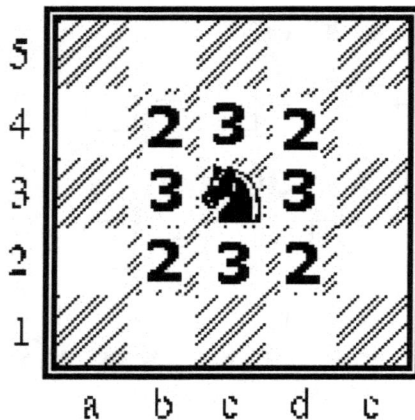

Look the number of moves it takes to reach b4 and c4, the distance to reach b4 and c4 do not reflect the real distance.

It would take longer to reach b4 than c4 in real distance.
How many squares does it take to reach e2 from Nc3? _____

Student's Name _____   Date

Direction, geometry translation, chess notation

1	2	3	4	5
6	7	8	9	10
11	12	13	14	15
16	17	18	19	20
21	22	23	24	25

$13 + ⤢ + ✳ = 32$

$13 - ✳ - ✳ = $ ____

$8 + ⤢ - ⤢ = $ ____

$12 + ⊥ - ⤢ = $ ____

$24 ÷ 8 + ⬌ + ⬌ = $ ____

1. Mixed calculations

♛	♛	3	♛	4
+ 1	+ 2	+ ♛	+ 2	+ ♘
☐☐	☐☐	☐☐	☐☐	☐☐

♘	9	♘	6	♘
+ ♛	+ ♘	+ 7	+ ♘	+ 8
☐☐	☐☐	☐☐	☐	☐☐

2. Replace the question mark with a shape.

?

226

Student's Name _____ Date _____

Mixed operations

Chess diagram

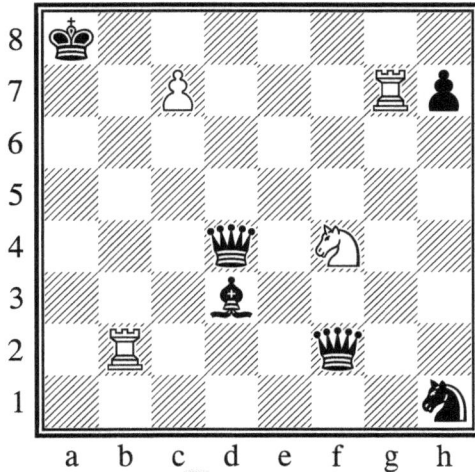

Fill in a number in each of □ , △, ○

The point of a square where ✳ can safely reach is □

□ - ○ = 4, ○= _____. ○ + ○ = △= _____ □ + △+○ = _____

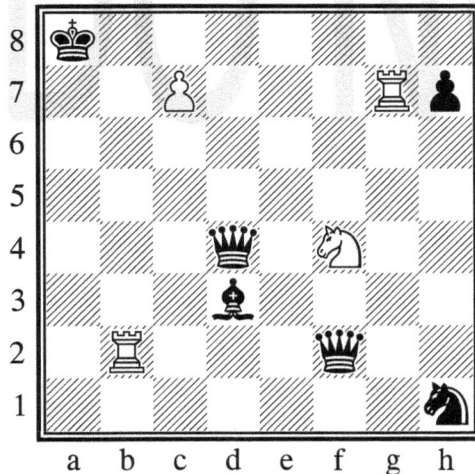

227

1. Mixed calculations

♕ + ♙ = 10   10 = ⬜ + ♙

♞ + 10 = ___   19 = ⬜ + ♕

♝ + 8 = ___   17 = ⬜ + ♖

♕ + ♖ = ___   14 = ⬜ + ♖

2. Complete the following diagram.

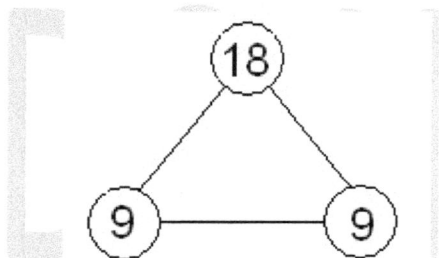

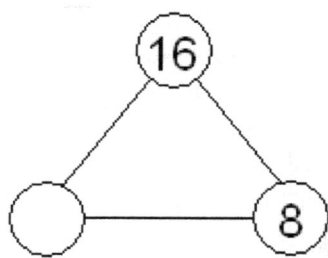

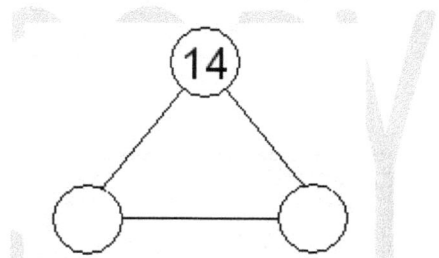

3. Fill in the blank.

⬜   ⬜   ⬜   ⬜   ⬜
+ 6   + ♝   + 8   + 7   + ♖
1 2   7   1 8   1 1   1 1

⬜   ⬜   ⬜   ⬜   ⬜
+ ♕   + ♞   + 7   + ♞   + 8
1 7   5   1 3   8   1 5

Student's Name _____   Date _____

Fill in 2 circles with 2 numbers having a difference of 2

Student's Name _____   Date _____

Fill in 2 circles with 2 numbers having a difference of 3.

Student's Name   Date

Mixed computations

Fill in 2 circles with 2 consecutive natural numbers having a difference of 1

Student's Name _____   Date _____

Mixed computations

1. Complete the following table.

2. Fill in the following ☐ with a number.

8   6   4   2   ♔   ☐   ♘   ♖   7   ♕

3. Fill in ☐ with an answer.

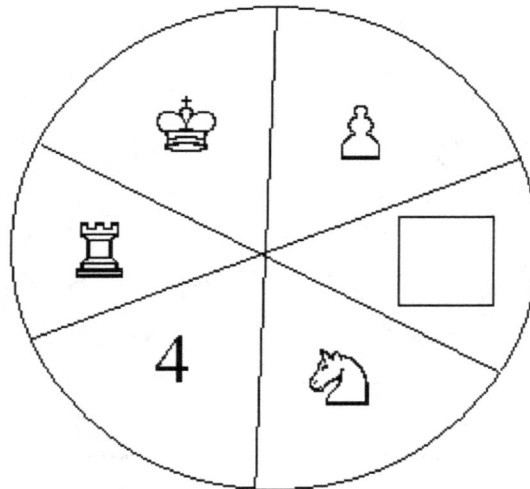

Fill in the following ☐ with a number, +, or – sign.

$$\square$$
$$+$$
$$9$$
$$\|$$
$$\square + 8 = 11 = 7 + \square$$
$$-\ 5$$
$$\square - 5 = \square = 11 - \square$$

$$\|$$
$$9$$
$$\square$$
$$2$$

$$\|$$
$$11$$
$$\square$$
$$6$$

Fill in the following $\square$ with a number, +, or – sign.

$$\square$$
$$+$$
$$9$$
$$=$$

$$\square + 8 = 10 = 7 + \square$$

$$-5$$

$$\square - 5 = \square = 10 - \square$$

$$=$$
$$9$$
$$\square$$
$$1$$

$$=$$
$$10$$
$$\square$$
$$5$$

何数棋谜　低年级棋谜式数学

Student's Name _____　　　Date _____

Fill in the following ☐ **with a number, +, or – sign.**

$$\square$$
$$+$$
$$9$$
$$\|$$
$$\square + 8 = 12 = 7 + \square$$
$$\underline{-\ 6}$$
$$\square - 6 = \square = 12 - \square$$
$$\|\qquad\qquad\qquad\qquad\|$$
$$8\qquad\qquad\qquad\qquad 12$$
$$\square\qquad\qquad\qquad\qquad \square$$
$$4\qquad\qquad\qquad\qquad 6$$

1. Mixed computations

♔ = 0, ♙ = 1, ♗ = 3, ♘ = 3, ♖ = 5, ♕ = 9

♖	6	7	♔	4
+ 0	+ ♕	+ ♘	+ 8	+ ♗
☐	☐☐	☐☐	☐	☐

♘	7	6	♕	5
+ 5	+ ♗	+ ♙	+ 0	+ ♖
☐	☐☐	☐	☐	☐☐

6	♖	♕	4	♔
+ ♘	+ 9	+ 3	+ ♗	+ 8
☐	☐☐	☐☐	☐	☐

2. Compare which chessboard (Black + White altogether) has more pawns by marking ">" (greater than), "<"(less than) or "="(equal to).

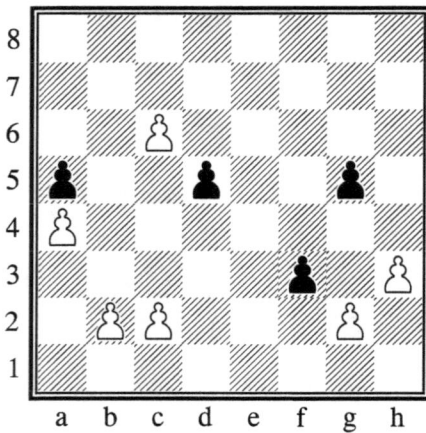

_____ pawns

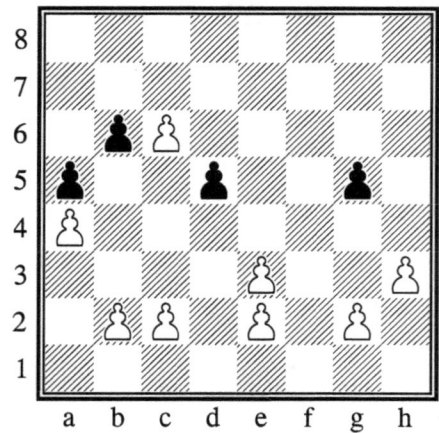

_____ pawns

Student's Name _____   Date _____

Mixed computation

You are at b2 = ☐.

3	5	3	2
2	8	**4**	4
1	7	9	6
	a	**b**	**c**

Mixed computations

You are at b2 = ☐.

3	1	2	3
2	8	1 3 / 7 5	4
1	7	6	5
	a	b	c

Student's Name ______________________ Date ______________

1. Mixed computations

♔ = 0, ♙ = 1, ♗ = 3, ♘ = 3, ♖ = 5, ♕ = 9

3 + 4 = 7 3 + 5 = _____

4 + 6 = _____ 5 + 6 = _____

7 + 5 = _____ 7 + 7 = _____

6 + 7 = _____ 7 + 9 = _____

8 + 6 = _____ 9 + 8 = _____

2. Complete the table.

♙	♘	♖	♘	♗
+ 6	+ 5	+ 7	+ 4	+ 5
□	□	□□	□	□
♙	♘	♖	♘	♗
+ 5	+ 3	+ 6	+ 4	+ 6
□	□	□□	□	□
♕	♘	♗	♕	♖
+ 2	+ 5	+ 7	+ 3	+ 7
□□	□	□□	□□	□□
♘	♗	♕	♖	♘
+ 7	+ 8	+ 4	+ 5	+ 8
□□	□□	□□	□□	□□

Student's Name _____ Date _____

Mixed computations

✳ ♔ = 0, ♛ ♙ = 1, ✳ ♗ = 3, ♘ ♞ = 3, ♘ ♖ = 5, ✳ ♕ = 9

Add the sum of numbers as directed.

3	4	5	7	9
1	♙	6	♙	6
4	♗	♕	♖	4
6	♗	♗	♖	5
7	6	8	4	3

$$\frac{5 + 6}{= 11}$$

3	4	5	7	9
1	5	♙	3	6
5	1	♗	♙	6
9	♗	♖	6	8
7	6	8	4	3

3	4	5	7	9
1	♗	5	♙	6
5	3	♖	4	5
3	♗	♖	♖	4
7	6	8	4	3

Student's Name _____ Date _____

Mixed computations

You are at b2 = ☐.

3	8	6	6
2	7	4 6 / 5 8	4
1	7	4	2
	a	b	c

Mixed computations

✳ ♔ = 0, ♕ ♙ = 1, ✕ ♗ = 3, ✚ ♘ = 3, ✛ ♖ = 5, ✳ ♛ = 9

1. Connect column 2 to match column 4.

♖ + ✛ + ♖ + ✛   8

♞ + ✚ + ♞   19

♗ + ♗ + ✕ + ♗   20

♛ + ✳   15

↓ + ♟ + ♟ + ↓ + ♟ + ↓ + ♟ + ↓   9

♚ + ♟ + ✳ + ♞ + ♟ + ↓   12

✚ + ♞ + ✚ + ♖ + ♖   18

2. Replace each ? with a chess piece.

♕	✕	✳	♗	♖
♞	♕	✕	✳	♗
♙	♞	?	✕	✳
✕	♙	♞	♕	✕
♖	✕	?	♞	♕

Student's Name _____　Date _____

Mixed computations

✳ ♔ = 0, ♙ ♙ = 1, ✕ ♗ = 3, ✛ ♘ = 3, ✛ ♖ = 5, ✳ ♕ = 9

Compute the following chess values according to the directions of knight moves.

$$\frac{}{} + \frac{}{}$$

= ___ + ___

= ___

$$+$$

= ___ + ___

= ___

$$+$$

= ___ + ___

= ___

$$+$$

= ___ + ___

= ___

Student's Name _____ Date _____

Mixed computations

♦♔ = 0, ♙♟ = 1, ✠♝ = 3, ✚♞ = 3, ✛♖ = 5, ✹♕ = 9

Fill in blanks.

♕ + ✠ 9 + 3 = 12	♖ + ✛ 5 + 5 = ___ 10
✛ +7 ___ + ___ = ___	✠ +10 ___ + ___ = ___
(♖ + ♞) +2 ___ + ___ = ___	(♞ + ♞) + (♞ + ♟) ___ + ___ = ___
(♙ + ✚) + ✠ ___ + ___ = ___	(♙ +5) + (✠ +4) ___ + ___ = ___
✹ + (✛ + ✠) ___ + ___ = ___	(3 + ✚) + (✠ +7) ___ + ___ = ___
(✛ + ♖) + (♞ + ✠) ___ + ___ = ___	(✹ + ♟) + (✛ + ✠) ___ + ___ = ___
✹ + (♖ + ✛) ___ + ___ = ___	(♙ + ✠) + (✛ + ✠) ___ + ___ = ___

Student's Name _____　　　　　　Date _____

9 + d, doubles, adding to 10, subtractions

```
  0 3 0 5 0 3 0 5 0 6 0 9 0 8 0 9 0 8 0 4 0 3
+ 0 9 0 7 0 9 0 7 0 6 0 3 0 4 0 3 0 4 0 8 0 9
```

```
  0 7 0 3 0 9 0 4 0 5 0 4 0 3 0 8 0 7 0 5 0 6
+ 0 5 0 9 0 3 0 8 0 7 0 8 0 9 0 4 0 5 0 7 0 6
```

```
  0 4 0 3 0 8 0 6 0 5 0 4 0 5 0 9 0 8 0 6 0 5
+ 0 8 0 9 0 4 0 6 0 7 0 8 0 7 0 3 0 4 0 6 0 7
```

```
  0 9 0 4 0 3 0 6 0 5 0 4 0 6 0 4 0 4 0 8 0 7
+ 0 3 0 8 0 9 0 6 0 7 0 8 0 6 0 8 0 8 0 4 0 5
```

```
  0   0   0   0   0   0   0   0   0   0   0
+ 0 4 0 6 0 5 0 7 0 4 0 3 0 7 0 3 0 8 0 3 0 4
  1 2 1 2 1 2 1 2 1 2 1 2 1 2 1 2 1 2 1 2 1 2
```

```
  0 6 0 7 0 6 0 8 0 6 0 7 0 6 0 6 0 6 0 6 0 7
+ 0   0   0   0   0   0   0   0   0   0   0
  1 2 1 2 1 2 1 2 1 2 1 2 1 2 1 2 1 2 1 2 1 2
```

Student's Name _____   Date _____

Spatial relation and addition operation of mixed computations

7	1	3
6	15	5
2	8	4

9	4	1
6	5	5
7	3	8

3	5	7
4	12	1
8	9	6

1	3	8
5	8	6
9	7	4

$\underline{15} - \underline{1} = \underline{14}$ $\underline{5} + \underline{4} = \underline{9}$ $\underline{12} - \underline{5} = \underline{7}$ $\underline{8} + \underline{3} = \underline{11}$

___ − ___ = ___   ___ + ___ = ___   ___ − ___ = ___   ___ + ___ = ___

___ − ___ = ___   ___ + ___ = ___   ___ − ___ = ___   ___ + ___ = ___

___ − ___ = ___   ___ + ___ = ___   ___ − ___ = ___   ___ + ___ = ___

___ − ___ = ___   ___ + ___ = ___   ___ − ___ = ___   ___ + ___ = ___

___ − ___ = ___   ___ + ___ = ___   ___ − ___ = ___   ___ + ___ = ___

___ − ___ = ___   ___ + ___ = ___   ___ − ___ = ___   ___ + ___ = ___

___ − ___ = ___   ___ + ___ = ___   ___ − ___ = ___   ___ + ___ = ___

Student's Name _____   Date _____

Spatial relation and addition operation of mixed computations

7	8	3
6	5	5
2	9	4

7	2	4
6	7	5
9	8	3

4	9	5
8	9	2
7	3	6

5	4	7
5	6	9
3	8	6

$5 + 8 = 13$ $7 + 2 = 9$ $9 + 9 = 18$ $6 + 4 = 10$

___ + ___ = ___   ___ + ___ = ___   ___ + ___ = ___   ___ + ___ = ___

___ + ___ = ___   ___ + ___ = ___   ___ + ___ = ___   ___ + ___ = ___

___ + ___ = ___   ___ + ___ = ___   ___ + ___ = ___   ___ + ___ = ___

___ + ___ = ___   ___ + ___ = ___   ___ + ___ = ___   ___ + ___ = ___

___ + ___ = ___   ___ + ___ = ___   ___ + ___ = ___   ___ + ___ = ___

___ + ___ = ___   ___ + ___ = ___   ___ + ___ = ___   ___ + ___ = ___

___ + ___ = ___   ___ + ___ = ___   ___ + ___ = ___   ___ + ___ = ___

Student's Name _____ Date _____

Spatial relation and addition operation of mixed computations

4	1	7
5	10	8
2	6	9

1	5	8
3	6	7
4	2	6

3	6	2
4	10	9
5	8	1

9	4	5
2	7	3
6	8	7

$\underline{10} - \underline{6} = \underline{4}$

$\underline{6} + \underline{2} = \underline{8}$

$\underline{10} - \underline{8} = \underline{2}$

$\underline{7} + \underline{8} = \underline{15}$

___ − ___ = ___   ___ + ___ = ___   ___ − ___ = ___   ___ + ___ = ___

___ − ___ = ___   ___ + ___ = ___   ___ − ___ = ___   ___ − ___ = ___

___ − ___ = ___   ___ + ___ = ___   ___ − ___ = ___   ___ + ___ = ___

___ − ___ = ___   ___ + ___ = ___   ___ − ___ = ___   ___ + ___ = ___

___ − ___ = ___   ___ + ___ = ___   ___ − ___ = ___   ___ + ___ = ___

___ − ___ = ___   ___ + ___ = ___   ___ − ___ = ___   ___ + ___ = ___

___ − ___ = ___   ___ + ___ = ___   ___ − ___ = ___   ___ + ___ = ___

Student's Name _____   Date _____

Mixed computations

♚ = 0, ♙ = 1, ♟ = 3, ♘ = 3, ♖ = 5, ♛ = 9

Fill in the following ☐ with a number.

$\updownarrow$ $+\quad 2$ ☐ $-\ 2 =$ ☐	2 $+\ $♖ ☐ $-\ \updownarrow =$ ☐
6 $+\ 1$ ☐ $-\ 1 =$ ☐	$\downarrow$ $+\quad 6$ ☐ $-\ 6 =$ ☐
✳ $+\ $✕ ☐ $-\ $✕ $=$ ☐	♔ ✳ $+$ ☐ $-\ 9 =$ ☐
3 $+\ 4$ ☐ $-\ 4 =$ ☐	4 $+\ 3$ ☐ $-\ 3 =$ ☐
4 $+\ $$\leftrightarrow$ ☐ $-\ $♖ $=$ ☐	♖ $+\ 4$ ☐ $-\ 4 =$ ☐
7 $+\ 6$ ☐ $-\ 6 =$ ☐	6 $+\ 7$ ☐ $-\ 7 =$ ☐

Student's Name _____   Date _____

Mixed computations

✳ ♔ = 0, ♛ ♙ = 1, ✖ ♝ = 3, ✚ ♞ = 3, ✚ ♖ = 5, ✳ ♕ = 9

Table	Direction	Computation
♙ ♞ ♟ / 7 4 6 / ✳ 8 0		____ + ____ = ____ ____ − ____ = ____
♟ 3 4 / 7 ♖ 7 / ♕ 8 ♚		____ + ____ = ____ ____ − ____ = ____
1 ♝ 4 / 7 6 4 / ♕ 8 ♞		____ + ____ = ____ ____ − ____ = ____
1 ✳ 4 / 7 9 6 / 6 ♝ ♞		____ + ____ = ____ ____ − ____ = ____

Student's Name _____   Date _____

Mixed computations

♔ = 0, ♙ = 1, ♝ = 3, ♞ = 3, ♖ = 5, ♛ = 9

Table	Direction	Computation
♙ ♞ ♞ / 7 4 6 / ✳ 8 0	✳	___ + ___ = ___ ___ − ___ = ___
♟ 2 4 / 7 ♖ 6 / ♛ 8 ♔	✳	___ + ___ = ___ ___ − ___ = ___
1 ♝ 4 / 7 6 3 / ♛ 8 ♞	✳	___ + ___ = ___ ___ − ___ = ___
1 ✳ 4 / 7 7 6 / 6 ♝ ♞	✳	___ + ___ = ___ ___ − ___ = ___

© 2007 – 2020   **Frank Ho, Amanda Ho**   All rights reserved. www.homathchess.com

Student's Name _____   Date _____

Mixed computations

❋ ♚ = 0, ♛ ♟ = 1, ✖ ♝ = 3, ✚ ♞ = 3, ✚ ♜ = 5, ❋ ♛ = 9

♚	♟	2
6	1	3
4	7	♛

A. _____ + _____ = _____

B. _____ − _____ = _____

0	1	2
♛	7	♞
4	8	4

C. _____ − _____ = _____

D. _____ + _____ = _____

♚	1	2	3	♞
6	♟	♝	4	3
4	7	♛	6	5
♞	♜	8	0	3
2	3	6	1	♟

E. _____ − _____ = _____

F. _____ + _____ = _____

3	♟	2	♝	♝
6	6	♞	4	♞
4	7	♛	6	♜
♞	♜	8	0	3
2	♝	6	2	♟

G. _____ − _____ = _____

H. _____ + _____ = _____

Student's Name _____   Date _____

Mixed computations

1. Reviewing

Count on top down	Count on bottom up	Borrow 10	Memorize	Test
8 + 3	3 + 8	11 − 8	11 − 3	11 − 8
6 + 4	4 + 6	10 − 4	10 − 6	10 − 4
8 + 5	5 + 8	13 − 5	13 − 8	13 − 5
7 + 4	4 + 7	11 − 4	11 − 7	11 − 7
8 + 6	6 + 8	14 − 6	14 − 8	14 − 8
9 + 4	4 + 9	13 − 4	13 − 9	13 − 4

computations

1. Calculating using chess pieces

1. ⚞ - ⛉ =	2. ⚜ - ⛉ =
3. ⚞ + ⛉ =	4. ⚜ + ⛉ =
5. ⚵ + ⛉ =	6. ⚵ - ⛉ =
7. ⚜ - ⛉ =	8. ✳ - ⛉ =
9. ✳ + ⛉ =	10. ⚵ + ⛉ + ⛉ =
11. ⚵ - ⛉ =	12. ⚵ + ⛉ =

2. How many more points is a rook than a pawn?

3. How many more points is a knight than a pawn?

4. How many more points is a queen than a pawn?

5. How many more points is a bishop than a pawn?

6. How many more points is a pawn than a king?

7. Which one of the chess pieces has the least point?

Student's Name _____   Date _____

Frankho ChessDoku, ChessMaze, and Math IQ Puzzles
何算独棋, 何谜宫棋, 智力数谜

Frankho ChessDoku™ 何算独棋

© 2008 Frank Ho, Amanda Yang

	a	b	c
3	1		
2		2	
1			

Ho Math and Chess

Number Puzzles 智力数谜
Find the number.

The same shape represents the same number. What number does each shape represent?

$$\square + \square - 2 = 4$$

$$\square = \_\_\_\_$$

Frankho ChessMaze 何谜宫棋

© 2008-2009   Frank Ho, Amanda Yang

Ho Math and Chess

Pattern 规律

What is the next one?

Student's Name _____   Date _____

Frankho ChessDoku™

© 2008 Frank Ho, Amanda Yang

	a	b	c
3	3		
2			2
1			

Ho **Math and Chess**

Find the number.

The same shape represents the same number. What number does each shape represent?

$$\bigcirc = 3$$

$$\Delta + \Delta + 2 = \Delta + \bigcirc$$

$$\Delta = \underline{\qquad}$$

Pattern

What comes next?

Frankho ChessMaze

© 2008-2009 Frank Ho, Amanda Yang

	a	b	c	d	e	f
6						♔
5						
4						
3						
2						
1	♕					

Ho **Math and Chess**

Student's Name _____   Date _____

Frankho ChessDoku™

© 2008 Frank Ho, Amanda Yang

Find the number.

The same shape represents the same number. What number does each shape represent?

$$\Diamond + \Diamond - 1 = 5$$

$$\Diamond = \underline{\hspace{2cm}}$$

Pattern

What comes next?

Frankho ChessMaze

© 2008-2009 Frank Ho, Amanda Yang

Student's Name _____   Date _____

Frankho ChessDoku™

© 2008 Frank Ho, Amanda Yang

	a	b	c
3	1	2	
2			
1			

Ho Math and Chess

Find the number.

The same shape represents the same number. What number does each shape represent?

$$2 + \diamondsuit = 2 - 0$$

$$\diamondsuit = \underline{\hspace{1cm}} 0$$

Frankho ChessMaze

© 2008-2009 Frank Ho, Amanda Yang

	a	b	c	d	e	f
6						♚
5						
4						
3						
2						
1	♛					

Ho Math and Chess

Pattern

What comes next?

Student's Name _____   Date _____

Frankho ChessDoku™

© 2008 Frank Ho, Amanda Yang

3			™
2	3		
1			1
	a	b	c

Ho Math and Chess

Find the number.

The same shape represents the same number. What number does each shape represent?

$$✩ + ✩ = 2$$

$$✩ + ✡ = 3$$

$$✩ = \rule{3cm}{0.4pt}$$

$$✡ = \rule{3cm}{0.4pt}$$

Frankho ChessMaze

Pattern

What comes next?

| • • • | ⋮ |
| ∴ | ? |

© 2008-2009 Frank Ho, Amanda Yang

6						♚
5						
4						
3						
2						
1	♛					
	a	b	c	d	e	f

Ho Math and Chess

Student's Name _____   Date _____

Frankho ChessDoku™

© 2008 Frank Ho, Amanda Yang

Ho Math and Chess

Find the number

The same shape represents the same number. What number does each shape represent?

$$9 + ✱ + ✱ + 1 = 16$$

$$✱ = \underline{\hspace{2cm}}$$

Pattern

What comes next?

Frankho ChessMaze

© 2008-2009 Frank Ho, Amanda Yang

Ho Math and Chess

260

Frankho ChessDoku™

© 2008 Frank Ho, Amanda Yang

	a	b	c
3	–	4	
2			
1		1	+

Ho Math and Chess

Find the number.

The same shape represents the same number. What number does each shape represent?

$$✡ + ✡ - 1 = 7$$

$$✡ = \underline{\hspace{2cm}}$$

Pattern

What comes next?

△　△

➡　?

Frankho ChessMaze

© 2008-2009　Frank Ho, Amanda Yang

	a	b	c	d	e	f
6						♔
5						
4						
3						
2						
1	♕					

Ho Math and Chess

Student's Name _____   Date _____

Frankho ChessDoku™

© 2008 Frank Ho, Amanda Yang

Ho Math and Chess

Find the number.

The same shape represents the same number. What number does each shape represent?

$$\square - 1 + \square - 1 = 2$$

$$\square = \underline{\hspace{3cm}}$$

Pattern

What comes next?

Frankho ChessMaze

© 2008-2009 Frank Ho, Amanda Yang

Ho Math and Chess

Frankho ChessDoku™

© 2008 Frank Ho, Amanda Yang

Find the number.

The same shape represents the same number. What number does each shape represent?

$$9 + 2 + 8 + 1 = ⛉ + ⛉$$

$$⛉ = \rule{3cm}{0.4pt}$$

Pattern

What comes next?

Frankho ChessMaze

© 2008-2009 Frank Ho, Amanda Yang

Student's Name _____ Date _____

Frankho ChessDoku™

© 2008 Frank Ho, Amanda Yang

	a	b	c
3		+	™
2			2
1	+	3	

Ho Math and Chess

Find the number.

The same shape represents the same number. What number does each shape represent?

$$9 = ☆ + ◇$$

$$☆ + 1 = ◇$$

$$☆ = \rule{1cm}{0.4pt} \quad ◇ = \rule{1cm}{0.4pt}$$

Frankho ChessMaze

© 2008-2009   Frank Ho, Amanda Yang

	a	b	c	d	e	f
6						
5						
4						
3						
2						
1						

Ho Math and Chess

Pattern

What comes next?

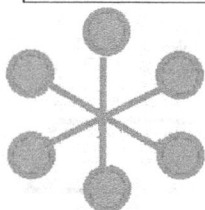

Student's Name _____   Date

Frankho ChessDoku™

© 2008 Frank Ho, Amanda Yang

Ho Math and Chess

Find the number.

The same shape represents the same number. What number does each shape represent?

$$3 + 1 = \square + 2$$

$$\square = \underline{\qquad}$$

Pattern

What comes next?

Frankho ChessMaze

© 2008-2009 Frank Ho, Amanda Yang

Ho Math and Chess

Frank Ho, Amanda Ho
Student's Name _____   Date _____

Frankho ChessDoku™

	a	b	c
3	1		™
2			
1	+	4	

Ho Math and Chess

Find the number.

The same shape represents the same number. What number does each shape represent?

$$7 + \Diamond = 6 + \square$$

$$\Diamond + \square = 3$$

$$\Diamond = \underline{\quad} \quad \square = \underline{\quad}$$

Frankho ChessMaze

	a	b	c	d	e	f
6						♔
5						
4						
3						
2						
1	♛					

Ho Math and Chess

Pattern

What comes next?

?

Student's Name _____　　Date _____

Frankho ChessDoku™

© 2008 Frank Ho, Amanda Yang

Ho Math and Chess

Find the number.

The same shape represents the same number. What number does each shape represent?

$$5 - \Diamond = 1 + \Diamond$$

$$\Diamond = \underline{\hspace{3cm}}$$

Pattern

What comes next?

Frankho ChessMaze

© 2008-2009 Frank Ho, Amanda Yang

Ho Math and Chess

Student's Name _____   Date _____

Frankho ChessDoku™

© 2008 Frank Ho, Amanda Yang

	a	b	c
3		3	
2	4		
1			

Ho **Math and Chess**

Find the number.

The same shape represents the same number. What number does each shape represent?

$$2 - \diamond = \diamond$$

$$\diamond = \rule{3cm}{0.4pt}$$

Frankho ChessMaze

© 2008-2009 Frank Ho, Amanda Yang

	a	b	c	d	e	f
6						
5						
4						
3						
2						
1						

Ho **Math and Chess**

Pattern

What comes next?

★	☆
▢	?

Student's Name _____   Date _____

Frankho ChessDoku™

© 2008 Frank Ho, Amanda Yang

Find the number.

The same shape represents the same number. What number does each shape represent?

$$5 - \diamondsuit = \diamondsuit + 3$$

$$\diamondsuit = \underline{\hspace{2cm}}$$

Frankho ChessMaze

Pattern

What comes next?

© 2008-2009   Frank Ho, Amanda Yang

Student's Name _____　　Date _____

Frankho ChessDoku™

© 2008　Frank Ho, Amanda Yang

Ho Math and Chess

Find the number.

The same shape represents the same number. What number does each shape represent?

$$4 - \text{☆} = 2 + \text{☆}$$

$$\text{☆} = \underline{\hspace{3cm}}$$

Pattern

What comes next?

Frankho ChessMaze

© 2008-2009　Frank Ho, Amanda Yang

Ho Math and Chess

270

Student's Name _____   Date _____

Frankho ChessDoku™

© 2008 Frank Ho, Amanda Yang

	a	b	c
3			
2			
1			

Ho Math and Chess

Find the number.

The same shape represents the same number. What number does each shape represent?

$$9 - \text{✦} = 3 + \text{✦}$$

$$\text{✦} = \underline{\qquad}$$

Frankho ChessMaze

Pattern

What comes next?

© 2008-2009 Frank Ho, Amanda Yang

	a	b	c	d	e	f
6						
5						
4						
3						
2						
1						

Ho Math and Chess

E up

271

Student's Name _____ Date _____

Frankho ChessDoku™

© 2008 Frank Ho, Amanda Yang

Ho Math and Chess

Find the number.

The same shape represents the same number. What number does each shape represent?

$$8 - \square = 2 + \square$$

$$\square = \underline{\hspace{2cm}}$$

Frankho ChessMaze

© 2008-2009 Frank Ho, Amanda Yang

Ho Math and Chess

Pattern

What comes next?

Student's Name _____　　Date _____

Frankho ChessDoku™

© 2008 Frank Ho, Amanda Yang

Ho Math and Chess

Find the number.

The same shape represents the same number. What number does each shape represent?

$$3 - ☆ = 1 + ☆$$

$$☆ = \_\_\_\_\_$$

Frankho ChessMaze

© 2008-2009 Frank Ho, Amanda Yang

Ho Math and Chess

Pattern

What comes next?

273

Student's Name _____   Date

Frankho ChessDoku™

© 2008 Frank Ho, Amanda Yang

	a	b	c
3		+	™
2	+		6
1		2	

Ho Math and Chess

Find the number.

The same shape represents the same number. What number does each shape represent?

$$8 - ☆ = ☆ + 4$$

$$☆ = \_\_\_\_$$

Frankho ChessMaze

© 2008-2009 Frank Ho, Amanda Yang

Ho Math and Chess

Pattern

What comes next?

?

Student's Name _____   Date _____

Frankho ChessDoku™

© 2008 Frank Ho, Amanda Yang

Ho Math and Chess

Find the number.

The same shape represents the same number. What number does each shape represent?

$$7 - \bigcirc = 1 + \bigcirc$$

$$\bigcirc = \underline{\hspace{2cm}}$$

Frankho ChessMaze

© 2008-2009 Frank Ho, Amanda Yang

Ho Math and Chess

Move diagonally.

Pattern

What comes next?

Student's Name   Date

Frankho ChessDoku™

© 2008 Frank Ho, Amanda Yang

Ho Math and Chess

Find the number.

The same shape represents the same number. What number does each shape represent?

$$6 - ⋈ = 2 + ⋈$$

$$⋈ = \underline{\hspace{2cm}}$$

Pattern

What comes next?

Frankho ChessMaze

© 2008-2009 Frank Ho, Amanda Yang

Ho Math and Chess

© 2007 — 2020   Frank Ho, Amanda Ho   All rights reserved. www.homathchess.com
Student's Name   Date

Frankho ChessDoku™

©2008 Frank Ho, Amanda Yang

Ho Math and Chess

Find the number.

The same shape represents the same number. What number does each shape represent?

$$7 = \square + \square - 1$$

$$\square = \underline{\qquad}$$

Pattern

What comes next?

Frankho ChessMaze

©2008-2009 Frank Ho, Amanda Yang

Ho Math and Chess

Student's Name _____   Date _____

Frankho ChessDoku™

© 2008 Frank Ho, Amanda Yang

Ho Math and Chess

Find the number.

The same shape represents the same number. What number does each shape represent?

$$\square + \text{✡} = 3$$

$$\square - \text{✡} = 1$$

$$\square = \underline{\hspace{2cm}}$$

$$\text{✡} = \underline{\hspace{2cm}}$$

Pattern

What comes next?

Frankho ChessMaze

© 2008-2009 Frank Ho, Amanda Yang

Ho Math and Chess

Math, Chess, and IQ Puzzles for the creative minds

只见棋谜不见题 劝君迷路不哭涕 数学象棋加谜题 健脑思维眞神奇
数学腦力棋力 三合一多功能 靜心又增智力 神奇創新教材

Teachers should decide to skip those puzzles which are too difficult for your students.

Amandaho Moving Dots Puzzle ™

You are at c3.

Move some dots in b3, c4, c2, or d3 squares into c3 square such that the sum of dots + dots in each of rook's moves at c3 will be equal to the number shown on its destination square. See the following example.

Example

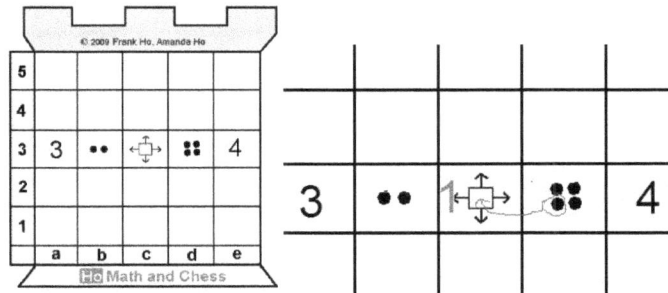

Problem

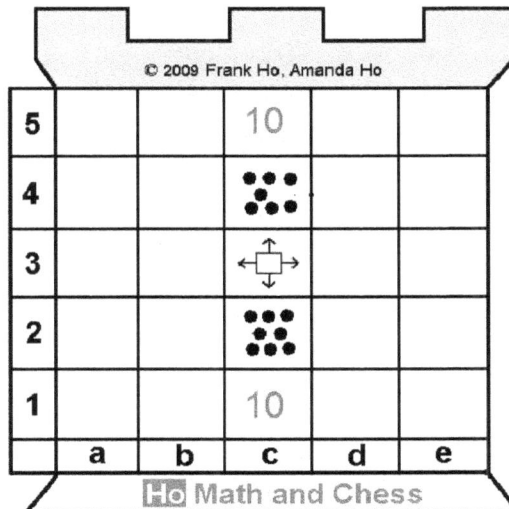

Student's Name _____ Date _____

Unequal Sudoku

Every row and column must have only one number starting from 1 to the number of squares of each side (Sudoku) but all numbers must obey the inequality sign.

Example

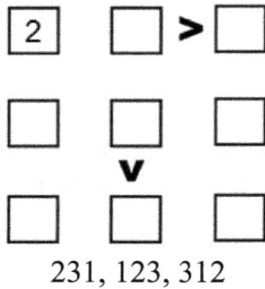

231, 123, 312

Problem

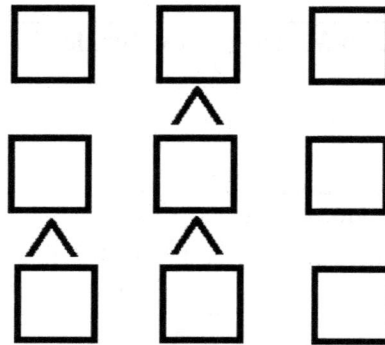

Loopy fence

Connect lines around each dot in such a way that each number indicates how many lines, connected by 4 dots only, surround it. The connected lines must form a single loop (like one rubber band) without lines crossed to each other.

Example

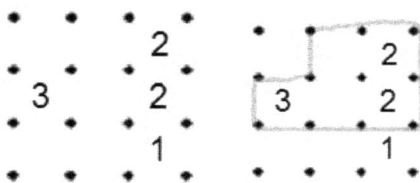

Problem

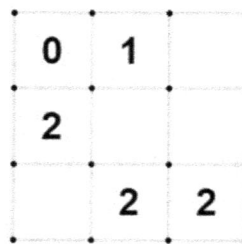

Student's Name _____　　　　Date _____

只见棋谜不见题　劝君迷路不哭涕　数学象棋加谜题　健脑思维眞神奇

Matchstick arithmetic

0 to 9 digits can be written by using matchsticks (without heads) as follows:

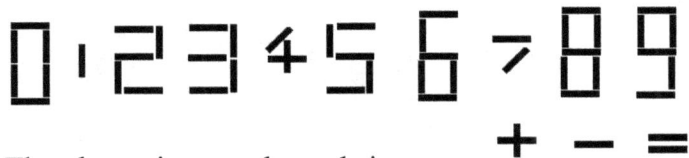

$$0123456789$$

$$+ - =$$

The plus, minus, and equal signs are:

A. Move one matchstick to make the following equation true.

$$2 + 1 = 4$$

B. . Move one matchstick to make the following equation true.

$$9 + 1 = 7$$

Student's Name _____   Date _____

The following chess diagram gives a value of 6.

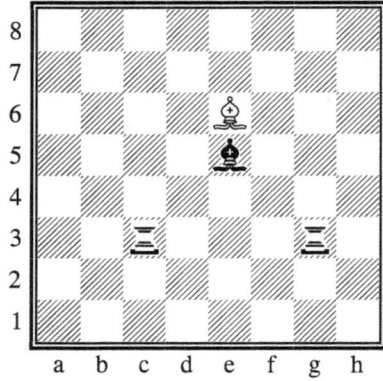

White bishop and 2 Black rooks intersect at 6 squares, so the value is 6.

What chess value shall the following chess diagram give?

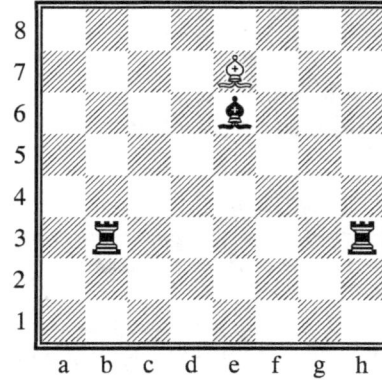

Observe each stickman and its pattern then fill in the empty circle with a number.

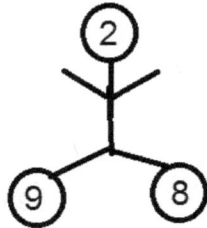

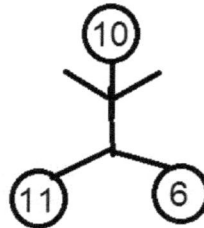

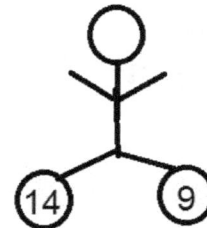

Student's Name _____   Date _____

Frankho Rook Path Puzzle	Draw line segments to show each rook's path.

Each rook's path is shown by either a vertical line **|** or horizontal line ▬ with the total number of vertical lines and horizontal lines equal to the number shown in the middle of each rook. No lines shall be crossed.

Example

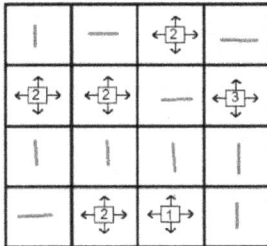

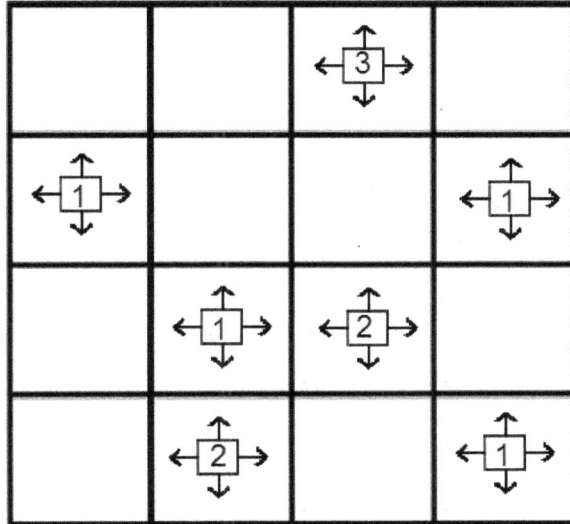

Student's Name _____   Date _____

Math and Chess Puzzle

You are at c3.

The points of chess pieces are as follows:

	a	b	c	d	e
5		⁘	✳	⁙	
4	∴	⊞	4	⤢	⁘
3	✳	4		2	✳
2	•	2	5	↓	∴
1		•	✳	∴	

✳ = 0

↓ = 1

⤢ = 3

⊞ = 3

⬌ = 5

✳✳ = 9

___ + ___ = ___ + ___ = ___

___ + ___ = ___ + ___ = ___

___ + ___ = ___ + ___ = ___

___ + ___ = ___ + ___ = ___

___ + ___ = ___ + ___ = ___

___ + ___ = ___ + ___ = ___

___ + ___ = ___ + ___ = ___

___ + ___ = ___ + ___ = ___

___ + ___ = ___ + ___ = ___

___ + ___ = ___ + ___ = ___

Student's Name _____　　　　　　Date _____

Amandaho Moving Dots Puzzle ™

You are at c3.

Move some dots in b3, c4, c2, or d3 squares into c3 square such that the sum of dots + dots in each of rook's moves at c3 will be equal to the number shown on its destination square. See the following example.

Example

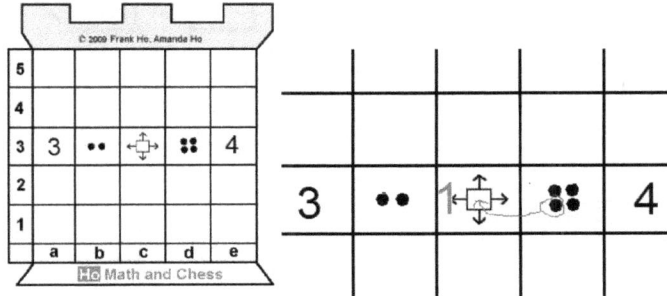

Problem

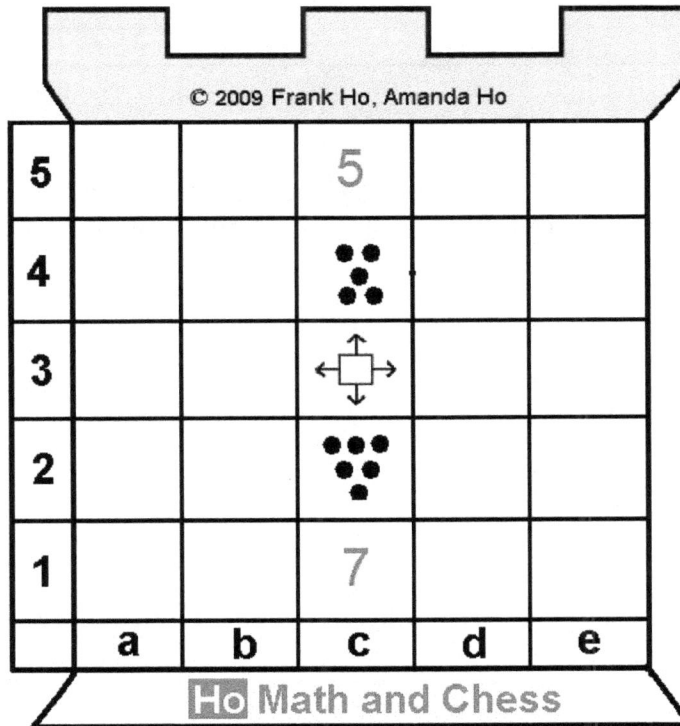

Unequal Sudoku

Every row and column must have only one number starting from 1 to the number of squares of each side (Sudoku) but all numbers must obey the inequality sign.

Example

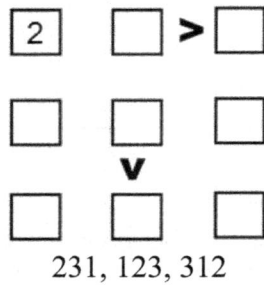

231, 123, 312

Problem

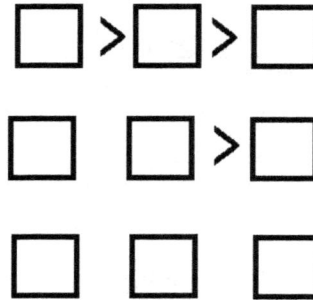

Loopy fence

Connect lines around each dot in such a way that each number indicates how many lines, connected by 4 dots only, surround it. The connected lines must form a single loop (like one rubber band) without lines crossed to each other.

Example

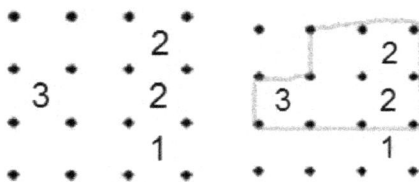

Problem

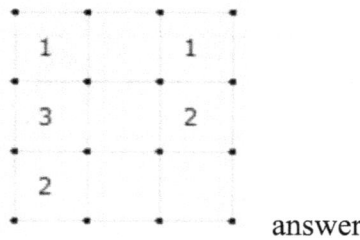

answer

Student's Name _____   Date _____

Frankho Rook Path Puzzle	Draw line segments to show each rook's path.

Each rook's path is shown by either a vertical line ❘ or horizontal line ▬ with the total number of vertical lines and horizontal lines equal to the number shown in the middle of each rook. No lines shall be crossed.

Example

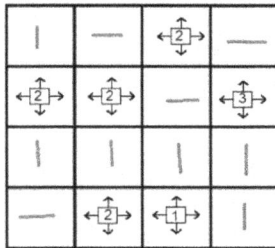

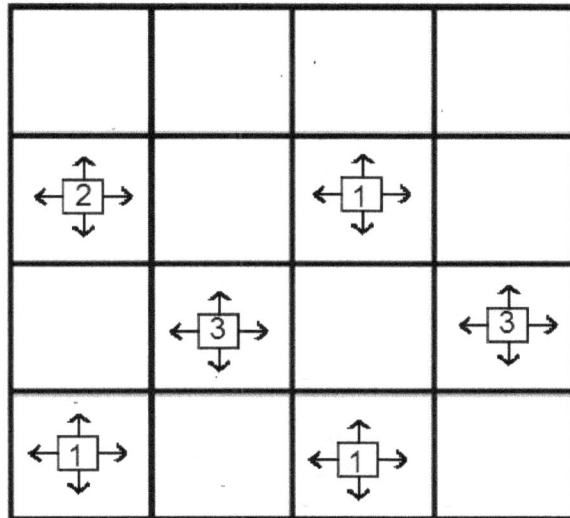

This problem is for advanced students only.

Find the price of apple, strawberry, banana, and pineapple.

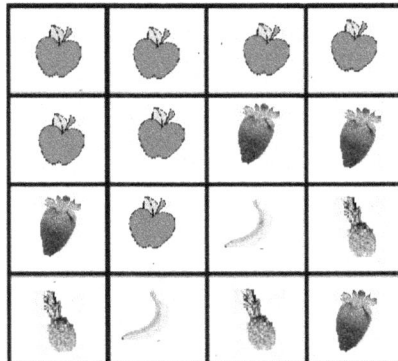

128 cents

130 cents

120 cents

127 cents

只见棋谜不见题 劝君迷路不哭涕 数学象棋加谜题 健脑思维眞神奇

数学脑力棋力 三合一多功能 静心又增智力 独一创新教材

No calculator, show all your work on empty space.

Bring these worksheets home; do not leave them in the classroom.

Lollipop stick arithmetic
0 to 9 digits can be written by using lollipop sticks as follows:

$$1\ 2\ 3\ 4\ 5\ 6\ 7\ 8\ 9\ 0$$

The plus, minus, and equal signs are $+\ -\ =$

Move 2 lollipop sticks such that the following equation is true.

$$4 + 2 - 1 = 8$$

Move 1 lollipop stick such that the following equation is true.

$$1 + 1 + 1 + 1 = 14$$

Move 1 lollipop stick such that the following equation is true.

$$14 + 13 - 7 = 0$$

Move 1 lollipop stick such that the following equation is true.

$$23 - 4 - 7 = 1$$

Student's Name _____　　　　Date _____

Lollipop stick figure

Move the following four lollipop sticks to make 3 squares.

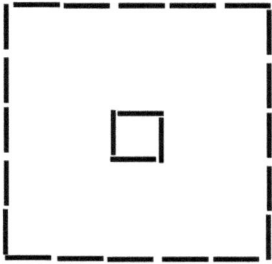

Move 2 lollipop sticks to get only 4 squares. No overlapping or loose ends.

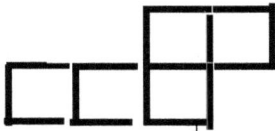

Observe each stickman and its pattern then fill in the empty circle with a number.

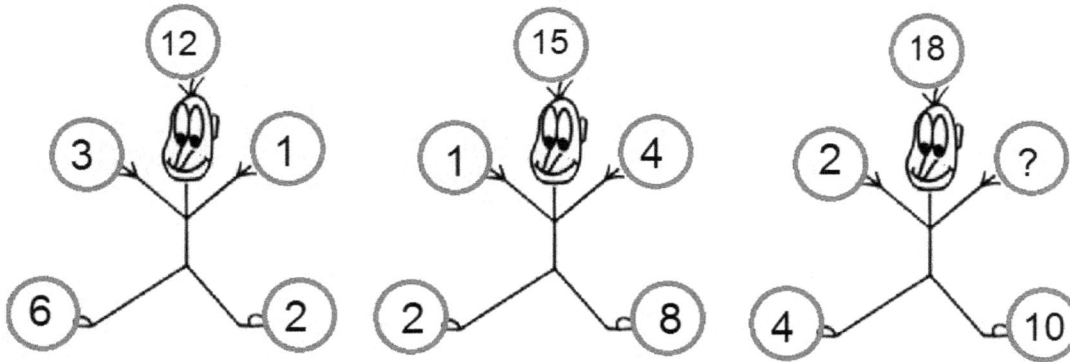

Student's Name _____   Date _____

Frankho Rook Path Puzzle	Draw line segments to show each rook's path.

Each rook's path is shown by either a vertical line **I** or horizontal line **▬** with the total number of vertical lines and horizontal lines equal to the number shown in the middle of each rook. No lines shall be crossed.

Example

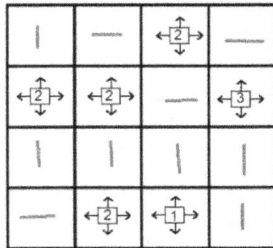

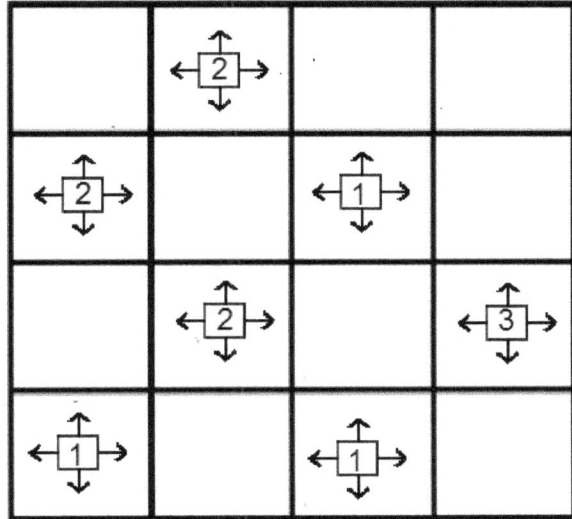

On the right figure, the numbers in the any two bottom adjacent circles add up to the number which is immediately above the bottom two circles. Fill only digits from 1 to 9 in each circle.

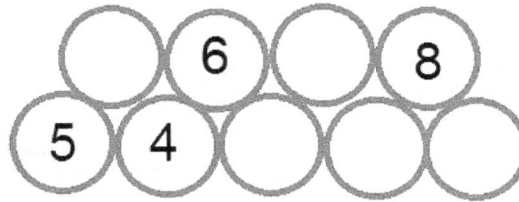

Student's Name _____   Date _____

Amandaho Moving Dots Puzzle ™

You are at c3.

Move some dots in b3, c4, c2, or d3 squares into c3 square such that the sum of dots + dots in each of rook's moves at c3 will be equal to the number shown on its destination square. See the following example.

Example

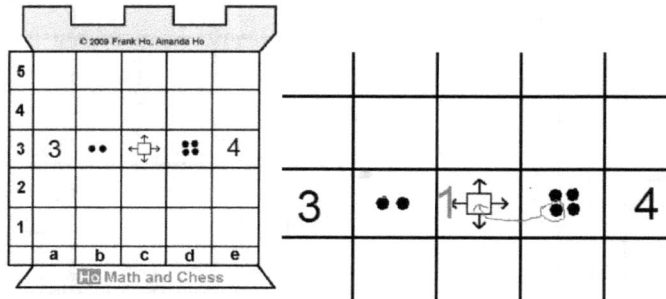

Problem

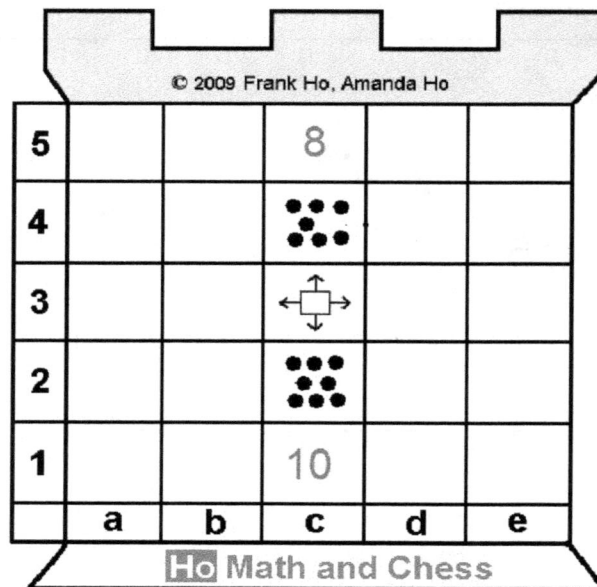

Student's Name _____   Date _____

Unequal Sudoku

Every row and column must have only one number starting from 1 to the number of squares of each side (Sudoku) but all numbers must obey the inequality sign.

Example

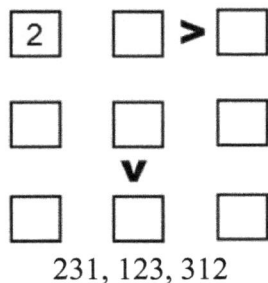

231, 123, 312

Problem

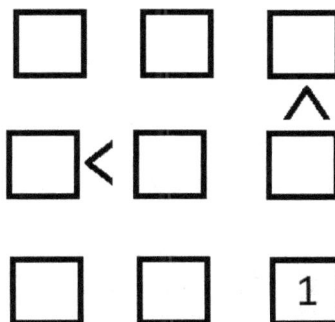

Loopy fence

Connect lines around each dot in such a way that each number indicates how many lines, connected by 4 dots only, surround it. The connected lines must form a single loop (like one rubber band) without lines crossed to each other.

Example

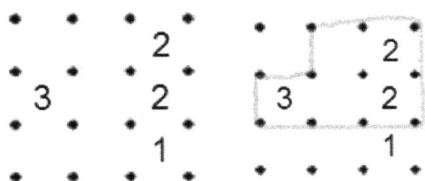

Problem

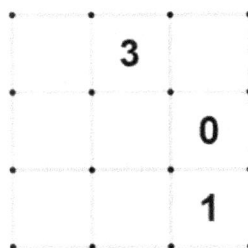

只见棋谜不见题   劝君迷路不哭涕   数学象棋加谜题   健脑思维真神奇

No calculator, show all your work on empty space.

Bring these worksheets home; do not leave them in the classroom.

1.

Matchstick arithmetic
0 to 9 digits can be written by using matchsticks (without heads) as follows:

0 1 2 3 4 5 6 7 8 9

The plus, minus, and equal signs are: + − =

C. Move one matchstick to make the following equation true.

1 + 1 = 3

D. . Move one matchstick to make the following equation true.

1 + 7 7 = 7

Student's Name _____   Date _____

| 2. | Matchstick figure

Move two matchsticks so that the fish below will change its current direction to a different direction. |

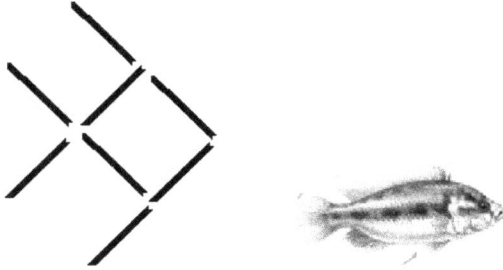

| 3. | The following chess diagram gives values of 5, 10, 10 when the bishop moves down. | What chess values shall the following chess diagram give when the bishop moves down? |

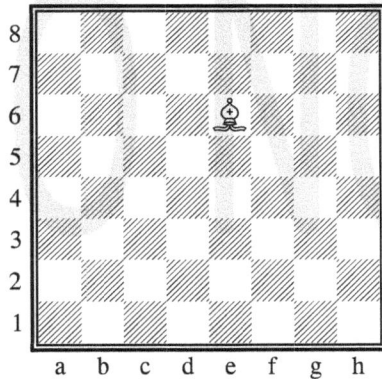

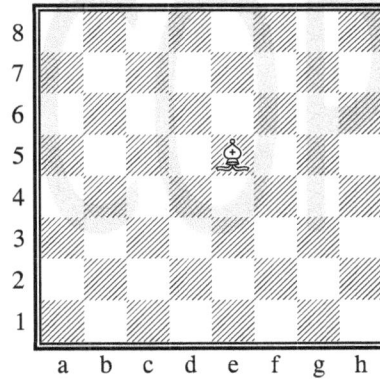

Student's Name _____   Date _____

4 | Observe each stickman and its pattern then fill in the empty circle with a number.

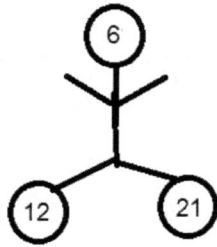

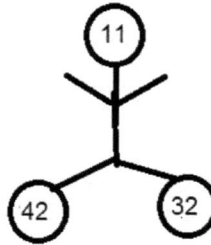

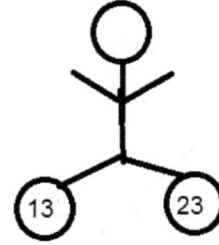

Student's Name _____   Date _____

Frankho Rook Path Puzzle	Draw line segments to show each rook's path.
Each rook's path is shown by either a vertical line **❙** or horizontal line **▬** with the total number of vertical lines and horizontal lines equal to the number shown in the middle of each rook. No lines shall be crossed. Example 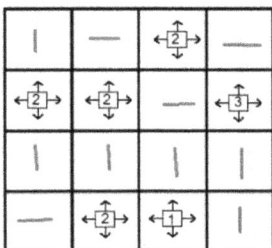	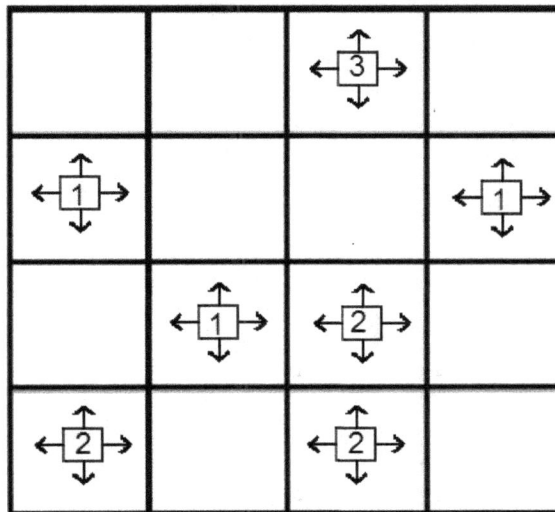

Replace each ? with a number.	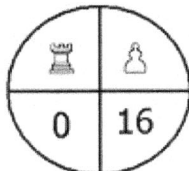	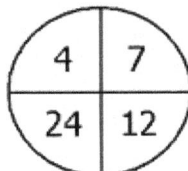	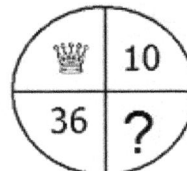

Student's Name _____   Date _____

Math and Chess Puzzle

You are at c3.

5		⠿	✳	⠿	
4	⠉	╪	5	✕	⠿
3	✳	9		7	✳
2	•	6	8	↓	⠿
1		•	✳	⠿	
	a	**b**	**c**	**d**	**e**

The points of chess pieces are as follows:

✳ = 0

↓ = 1

✕ = 3

╪ = 3

✛ = 5

✸ = 9

Student's Name _____   Date _____

Amandaho Moving Dots Puzzle ™

You are at c3.

Move some dots in b3, c4, c2, or d3 squares into c3 square such that the sum of dots + dots in each of rook's moves at c3 will be equal to the number shown on its destination square. See the following example.

Example

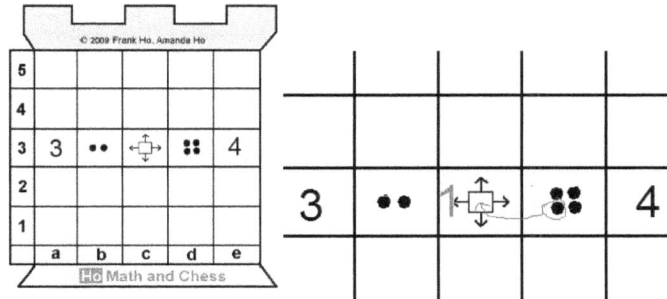

Problem

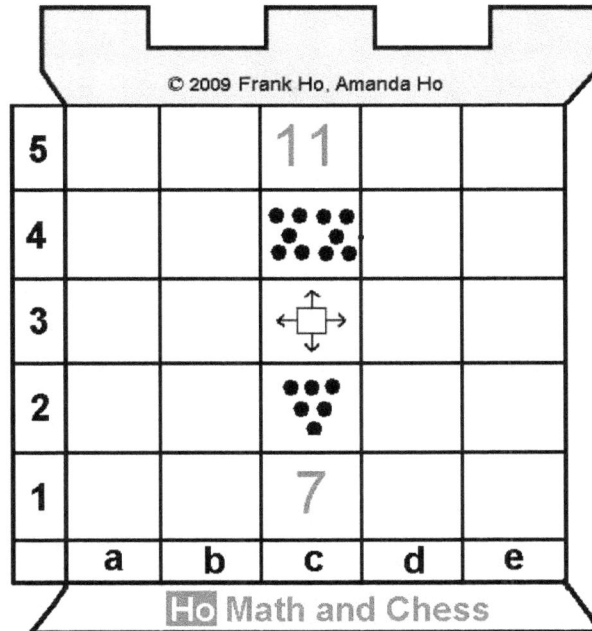

© 2007 – 2020　　Frank Ho, Amanda Ho　　All rights reserved. www.homathchess.com

Student's Name _____　　　　　Date _____

Unequal Sudoku

Every row and column must have only one number starting from 1 to the number of squares of each side (Sudoku) but all numbers must obey the inequality sign.

Example

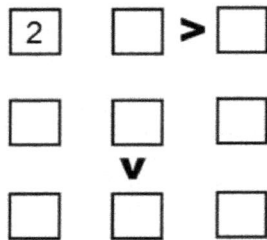

Problem

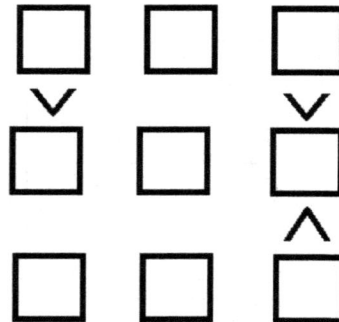

Loopy fence

Connect lines around each dot in such a way that each number indicates how many lines, connected by 4 dots only, surround it. The connected lines must form a single loop (like one rubber band) without lines crossed to each other.

Example

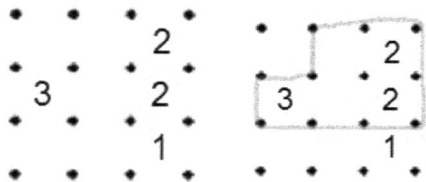

Problem

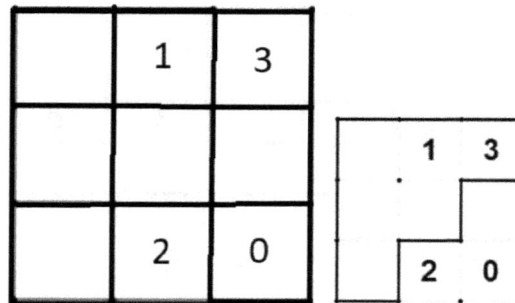

Student's Name _____　　Date _____

只见棋谜不见题　　劝君迷路不哭涕　　数学象棋加谜题　　健脑思维眞神奇

No calculator, show all your work on empty space.

1	Compute $11 + 12 + 13 + 14 - 1 - 2 - 3 - 4$

2	Circle the following odd one.

3	

Look at the left chess diagram and Replace the ? by a number.

13

9

3

10

10

?

10

4	Circle the odd one.

5	Replace each question by a digit. 310 (Change each chess piece by a number)

　???

Frankho Rook Path Puzzle

Each rook's path is shown by either a vertical line **|** or horizontal line **—** with the total number of vertical lines and horizontal lines equal to the number shown in the middle of each rook. No lines shall be crossed.

Example

Draw line segments to show each rook's path.

302

何数棋谜　低年级棋谜式数学

Student's Name _____　　Date _____

Amandaho Moving Dots Puzzle ™

You are at c3.

Move some dots in b3, c4, c2, or d3 squares into c3 square such that the sum of dots + dots in each of rook's moves at c3 will be equal to the number shown on its destination square. See the following example.

Example

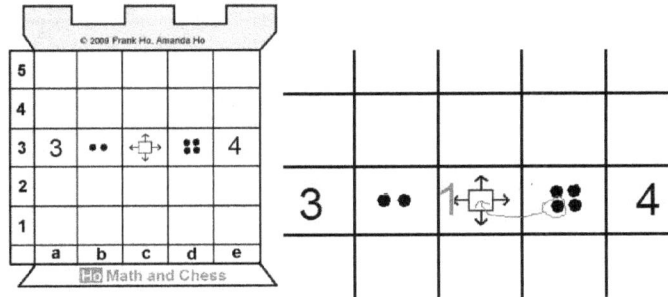

Problem

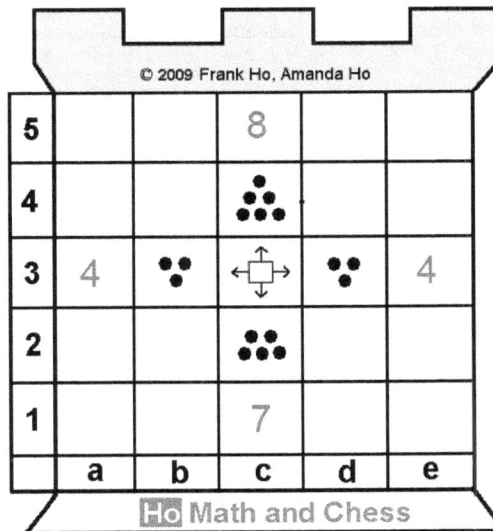

Unequal Sudoku

Every row and column must have only one number starting from 1 to the number of squares of each side (Sudoku) but all numbers must obey the inequality sign.

Example

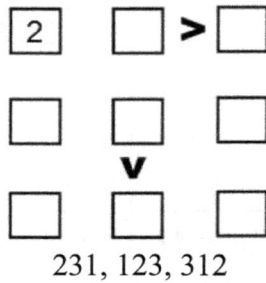

231, 123, 312

Problem

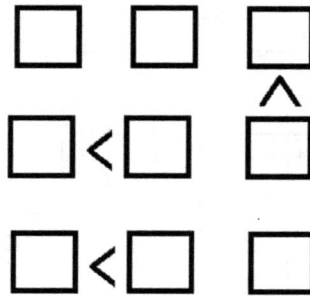

Loopy fence

Connect lines around each dot in such a way that each number indicates how many lines, connected by 4 dots only, surround it. The connected lines must form a single loop (like one rubber band) without lines crossed to each other.

Example

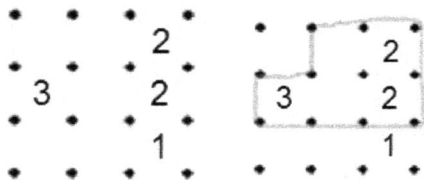

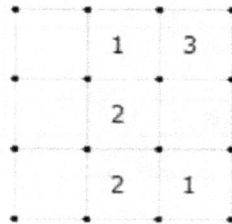

Problem

No calculator, show all your work on empty space.

Bring these worksheets home; do not leave them in the classroom.

Matchstick arithmetic
0 to 9 digits can be written by using matchsticks (without heads) as follows:

0 1 2 3 4 5 6 7 8 9

The plus, minus, and equal signs are: + – =

E. Move one matchstick to make the following equation true.

1 2 + 5 = 3

F. Move one matchstick such that the following four-digit number will be the largest.

1 9 9 5

G. Move one matchstick such that the following four-digit number will be the smallest. 4-digit.

1 9 9 5

Student's Name _____   Date _____

Matchstick figure

The following is upside down chair with one missing leg in the front. Move two matchsticks so that the chair will stand upright with no missing chair leg.

Matchstick figure

The following glass' opening toward up, move two matchsticks such that its opening will face down.

© 2007 − 2020   Frank Ho, Amanda Ho   All rights reserved. www.homathchess.com

Student's Name _____   Date _____

Fill in the missing 4 squares.

Observe each stickman and its pattern then fill in the empty circle with a number.

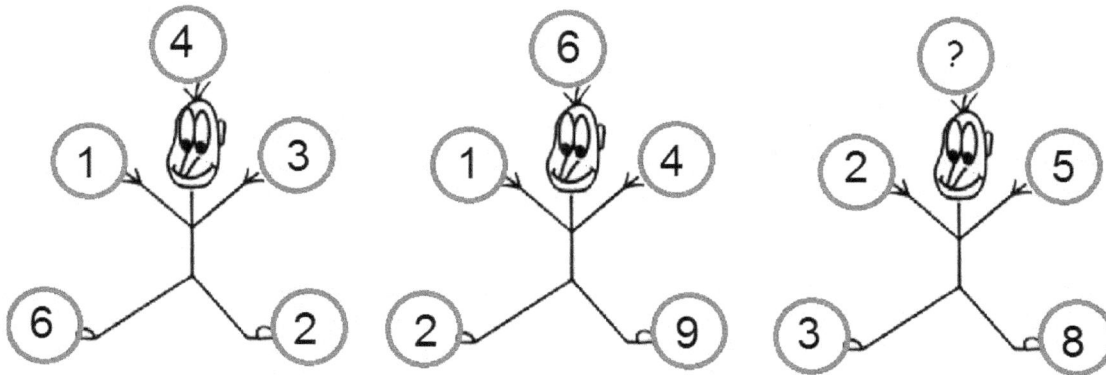

Student's Name _____　　Date _____

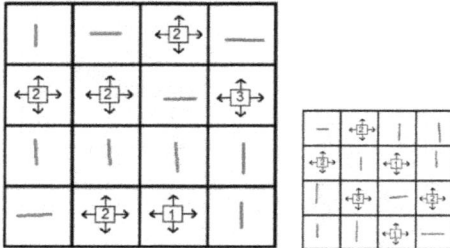

Frankho Rook Path Puzzle	Draw line segments to show each rook's path.

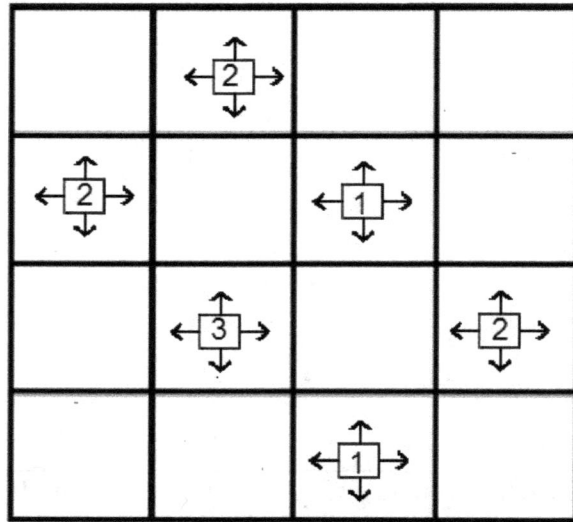

Each rook's path is shown by either a vertical line ▌ or horizontal line ▬ with the total number of vertical lines and horizontal lines equal to the number shown in the middle of each rook. No lines shall be crossed.

Example

Replace each ? with a number.

Amandaho Moving Dots Puzzle ™

You are at c3.

Move some dots in b3, c4, c2, or d3 squares into c3 square such that the sum of dots + dots in each of rook's moves at c3 will be equal to the number shown on its destination square. See the following example.

Example

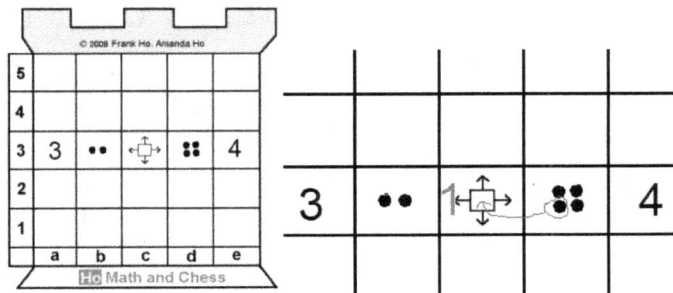

Problem

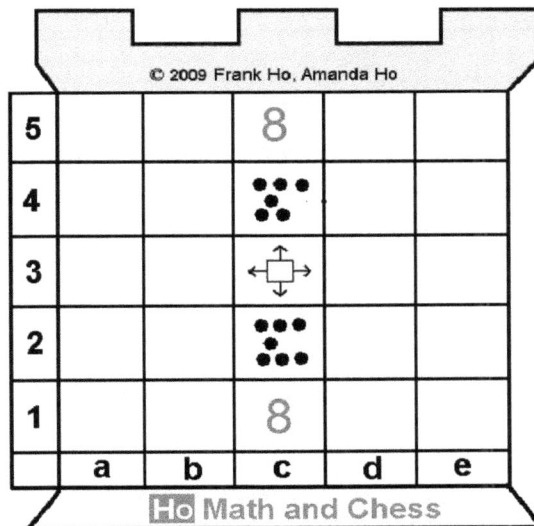

Unequal Sudoku

Every row and column must have only one number starting from 1 to the number of squares of each side (Sudoku) but all numbers must obey the inequality sign.

Example	Problem

Example:

2　□ > □

□　□　□
　　∨

□　□　□

231, 123, 312

Problem:

□ > □　□

□ < □　□
　　　　∨

□　□　□
　　　　∧

Loopy fence

Connect lines around each dot in such a way that each number indicates how many lines, connected by 4 dots only, surround it. The connected lines must form a single loop (like one rubber band) without lines crossed to each other.

Example	Problem

Example:

2
3　2
　　1

2
3　2
　　1

Problem:

2　0

2

3

只见棋谜不见题　劝君迷路不哭涕　数学象棋加谜题　健脑思维眞神奇
数学脑力棋力　三合一多功能　静心又增智力　神奇創新教材

No calculator, show all your work on empty space.

Bring these worksheets home; do not leave them in the classroom.

Lollipop stick arithmetic

0 to 9 digits can be written by using lollipop sticks as follows:

0 1 2 3 4 5 6 7 8 9

The plus, minus, and equal signs are: + – =

H. Move one matchstick to make the following equation true.

12 + 5 = 3

I. Move one matchstick such that the following four-digit number will be the largest.

1995

J. Move one matchstick such that the following four-digit number will be the smallest.
4-digit.

1995

Student's Name ____________________ Date ____________

Frankho ChessMaze and Castle Math

Trace the path from Queen to King.

Transformation	Symbol	Example
Slide	[-↔+ , ↕]	**[3, -2] Move right 3 squares and down 2 squartes.** **[0, 2] Move up 2 squares.**
Rotation	↱ ↰	↱✣ = ✣, ↰✣ = ✣
Flip	✣ flip	✣ flip = ✣, ✣ flip = ✣

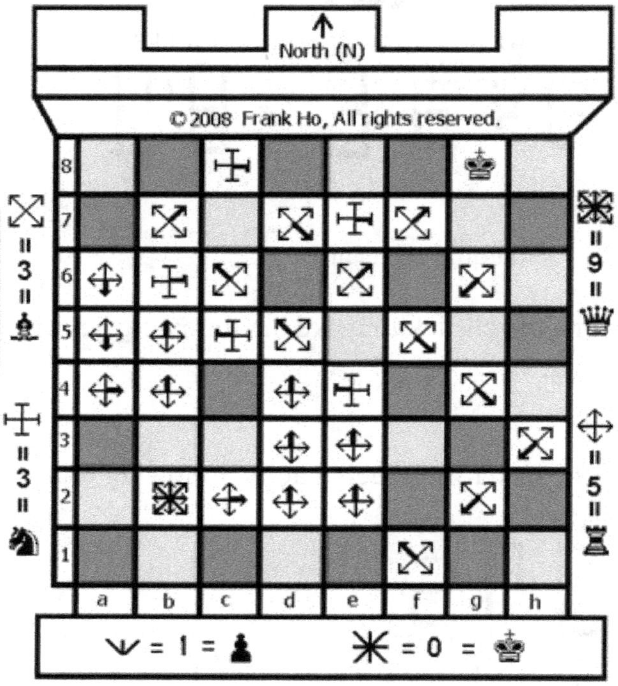
North (N)
© 2008 Frank Ho, All rights reserved.
8
7
6
5
4
3
2
1
a
b
c
d
e
f
g
h
= 3 =
= 3 =
= 9 =
= 5 =
= 1 =
= 0 =

Student's Name _____   Date _____

Frankho ChessMaze

Trace the path from Queen to King.

Transformation	Symbol	Example
Slide	[- ↔ + , ↕]	[3, -2] Move right 3 squares and down 2 squartes. [0, 2] Move up 2 squares.
Rotation	⌐→ ←¬	⟲ = ⇔ , ⟳ = ⇔
Flip	⇔ flip	⇔ flip = ⇔ , ⇕ flip = ⇔

⟱ = 1 = ♟   ✳ = 0 = ♚

Frankho ChessMaze

Trace the path from ⊠ to ♔. Movement direction is shown by a darker line segment.

Student's Name _____   Date _____

Frankho ChessMaze

Trace the path from ⊞ to ♔. Movement direction is shown by a darker line segment.

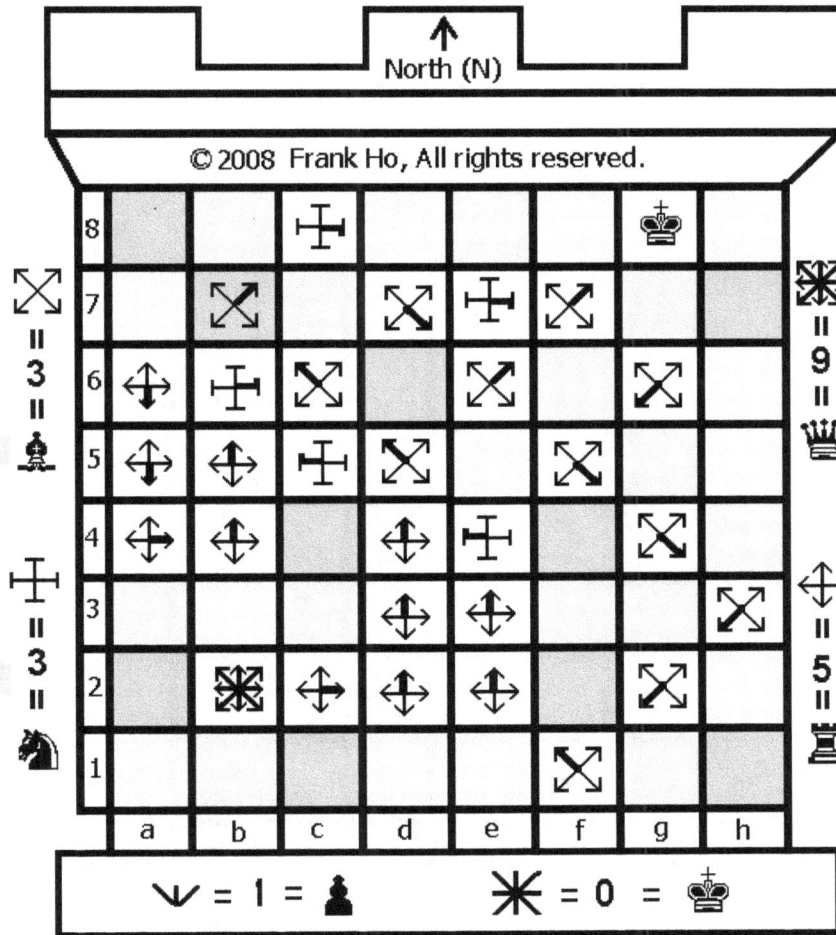

Frankho ChessMaze

Trace the path from ⊠ to ♚. Movement direction is shown by a darker line segment.

Frankho ChessMaze

Trace the path from ✳ to ♔. Movement direction is shown by a darker line segment.

Student's Name _____   Date _____

Frankho ChessMaze

1. Trace the path from ✳ to ♔.
2. Path direction is to move to a square with higher or equal value than the current one. Move one square at a time.

Transformation	Symbol	Example
Slide	$[-\leftrightarrow +, \updownarrow]$	[3, –2] Move right 3 squares and down 2 squartes. [0, 2] Move up 2 squares.
Rotation	⌐→ ⌐	🔄 = ⇔, 🔄 = ⇔
Flip	⇔ flip	⇔ flip = ⇔ , ⇕ flip = ⇔

It would be easier if students would just calculate the value of every square before making a move.	

Student's Name ______________________ Date ________

Frankho ChessMaze

Trace the path from Queen to King.

Transformation	Symbol	Example
Slide	[-↔+ , ↕]	[3, -2] Move right 3 squares and down 2 squartes. [0, 2] Move up 2 squares.
Rotation		
Flip	flip	flip = , flip =

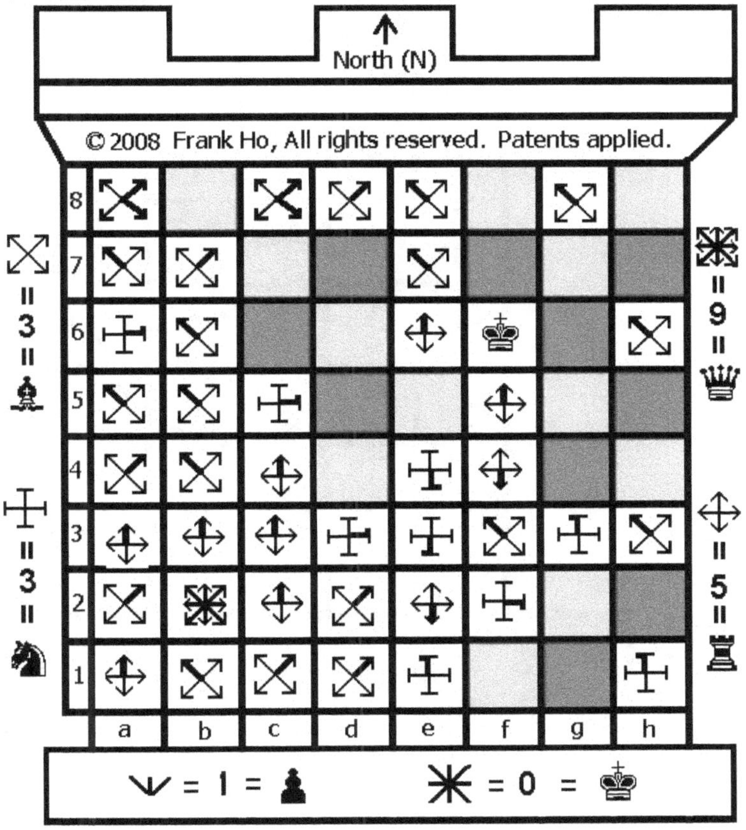

Frankho ChessMaze

Trace the path from ⊠ to ♔. Movement direction is shown by a darker line segment.

Frankho ChessMaze

Trace the path from ❋ to ♔. Movement direction is shown by a darker line segment.

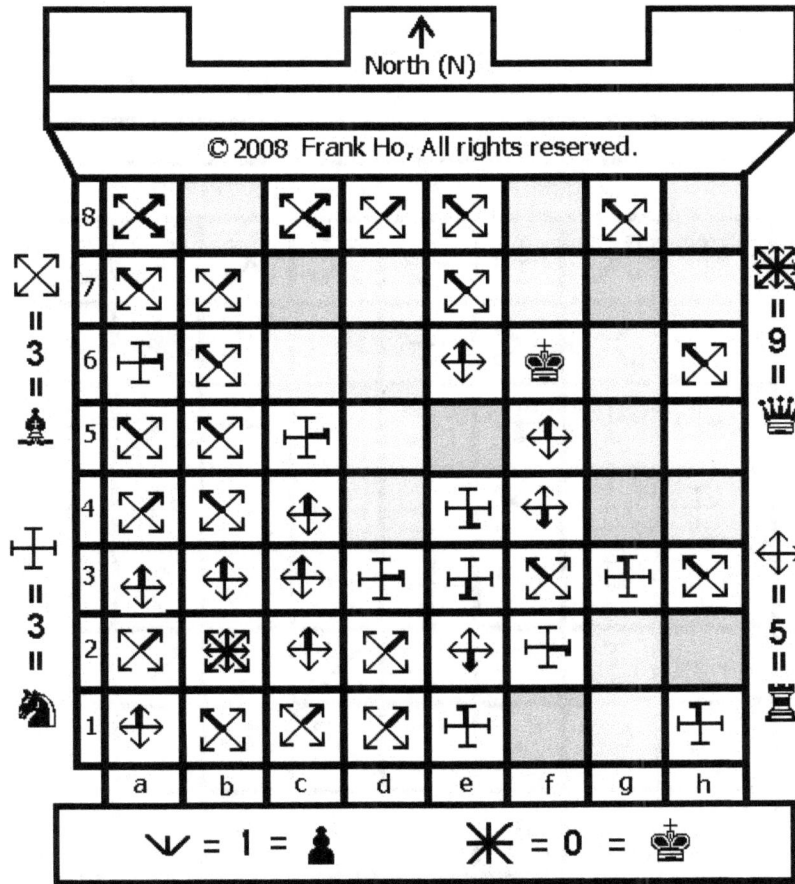

Student's Name _____   Date _____

Frankho ChessMaze

Trace the path from ⊞ to ♔. Movement direction is shown by a darker line segment.

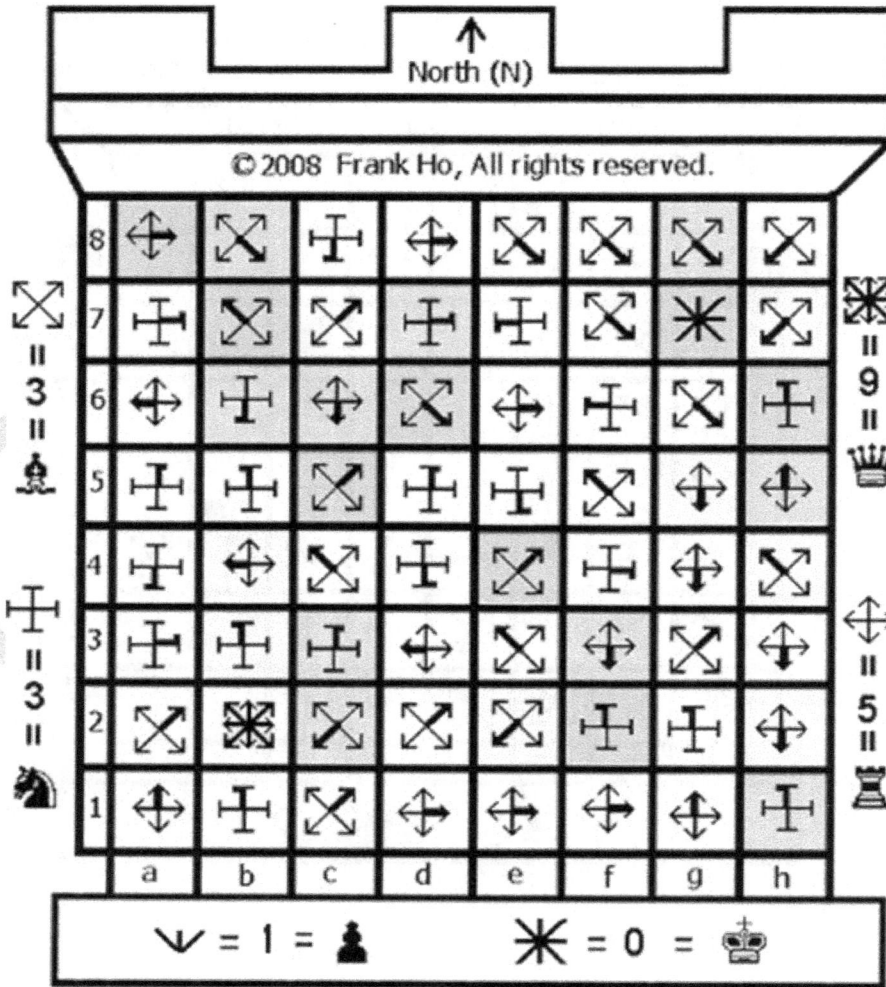

Student's Name _____   Date _____

Frankho ChessMaze

Trace the path from ✺ to ♔. Movement direction is shown by a darker line segment.

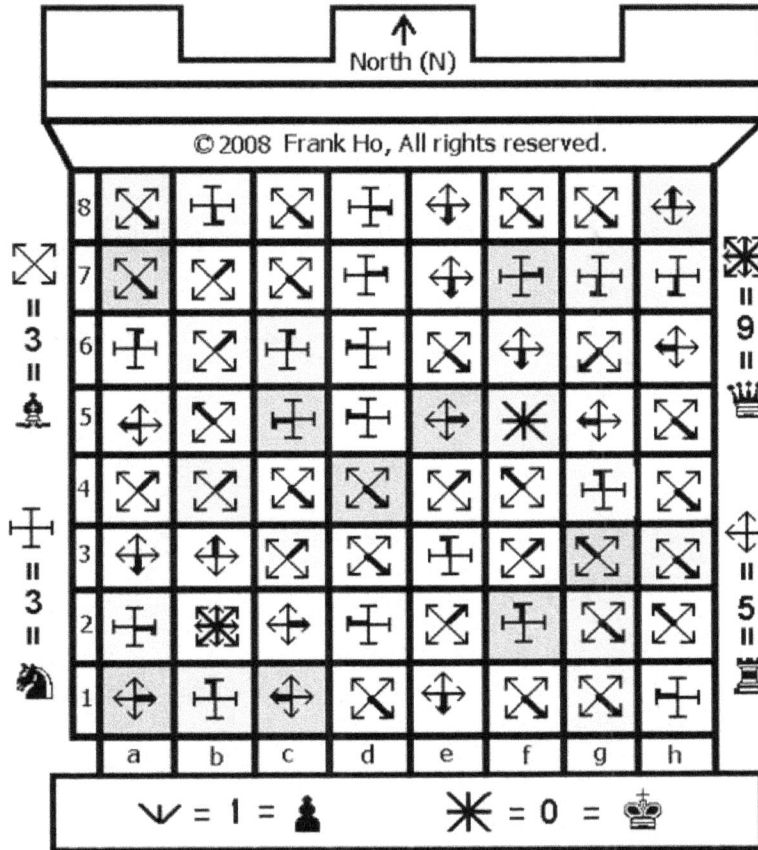

Frankho ChessMaze - trace and mark the shortest way for Ka1 to Kc10.

Student's Name _____ Date _____

Frankho ChessMaze - trace and mark the shortest way for Ke1 to Kg10.

Student's Name _____　　　　　　Date _____

Frankho ChessMaze

1. Trace the path from a8 to h1.
2. Move rook through squares with prime numbers. A prime number is a product of only 1 and itself.

Transformation	Symbol	Example
Slide	[-↔+ , ↕]	[3, –2] **Move right 3 squares and down 2 squartes.** [0, 2] **Move up 2 squares.**
Rotation	↱ ↰	⬚ = ⬚ , ⬚ = ⬚
Flip	⬚ flip	⬚ = ⬚ , ⬚ = ⬚

♚ = ✳ 　 North (N) 　 0 = ✳

©2008 Frank Ho, All rights reserved.

	a	b	c	d	e	f	g	h
8	♖	8	3	4	6	18	20	32
7	5	2	4	37	10	8	15	4
6	9	3	7	6	17	18	16	6
5	8	4	2	6	10	23	8	12
4	12	8	11	13	5	15	4	31
3	16	12	16	15	17	18	10	8
2	8	15	37	8	19	29	6	4
1	4	6	16	12	10	31	2	29

3 = ♞ = ⊥ = ♝ = ✕ , 5 = ♖ = ⬚ , 9 = ♛ = ✳

Frankho ChessMaze

1. Trace the path from ⊞ to ♔ □.
2. Movement direction is shown by a darker line segment.

Transformation	Symbol	Example
Slide	$[-\leftrightarrow+ , \updownarrow]$	[3, –2] Move right 3 squares and down 2 squartes. [0, 2] Move up 2 squares.
Rotation	⌐→ ←¬	⊞ = ⬌ , 卉 = ⬌
Flip	⬌ flip	⬌ = ⬌ , ⬌ = ⬌

PW (password):

Describe the slide rule how ▲ moves to △ using [___ , ___]

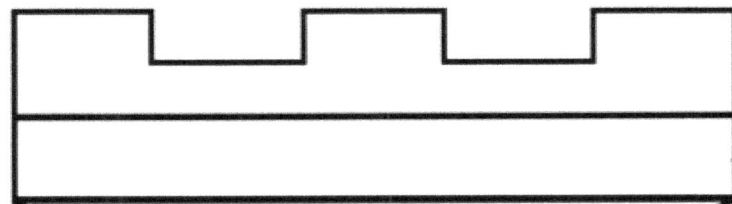

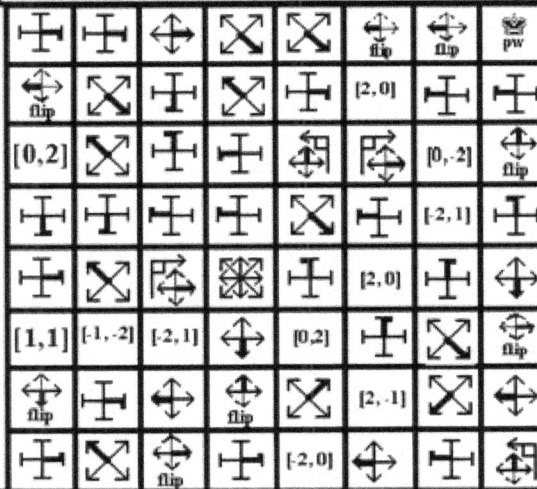

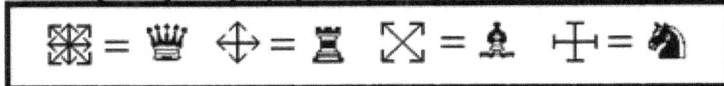

⊞ = ♛　⬌ = ♜　✕ = ♝　╬ = ♞

Maze

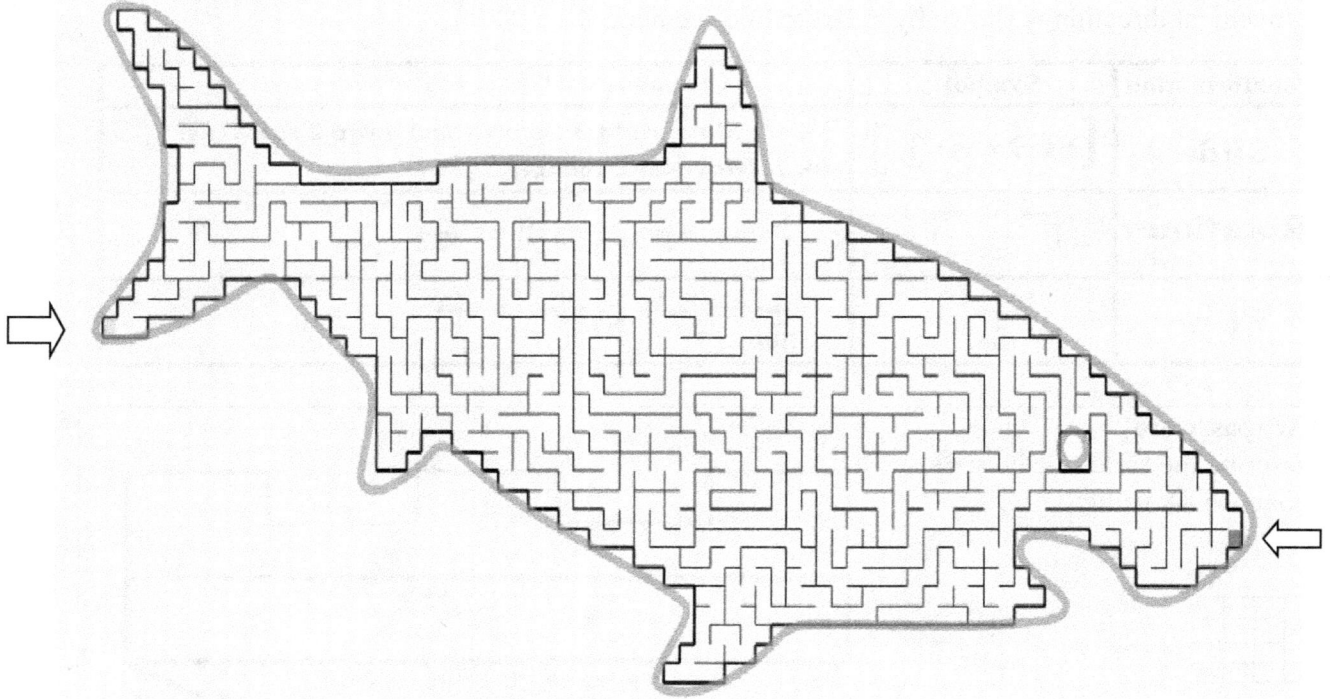

Ho Math Chess　何数棋谜　低年级棋谜式数学

© 2007 − 2020　　Frank Ho, Amanda Ho　　All rights reserved. www.homathchess.com

Student's Name　　　　　　　　　　　　　　　　　Date

Maze

Maze

Other Ho Math Chess student workbooks

Frank has written many workbooks for the elementary students. They are listed as follows:

1. Ultimate Math Contest Preparation, Problem Solving Strategies, and Math IQ Puzzles (3 in 1 workbook) Grade 1 and 2 (student 6717980, teacher 6720849)

2. Ultimate Math Contest Preparation, Problem Solving Strategies, and Math IQ Puzzles (3 in 1 workbook) Grade 2 and 3 (student 6745934, teacher 6783062)

3. Ultimate Math Contest Preparation, Problem Solving Strategies, and Math IQ Puzzles (3 in 1 workbook) Grade 3 and 4 (student 6812631, teacher 6813413)

4. Ultimate Math Contest Preparation, Problem Solving Strategies, and Math IQ Puzzles (3 in 1 workbook) Grade 4 and 5 (student 6906686, teacher 6907410)

5. Ultimate Math Contest Preparation, Problem Solving Strategies, and Math IQ Puzzles (3 in 1 workbook) Grade 5 and 6 (student 6979530, teacher 6985906)

6. Ultimate Math Contest Preparation, Problem Solving Strategies, and Math IQ Puzzles (3 in 1 workbook) Grade 6 and 7 (student 7060502, teacher 7060972)

7. Primary Grades Math (Grades 4 and under)

8. Elementary Grades Math (Grades 5 and Up)

9. Ho Math, Chess, and Puzzles for Grade 1 and Under

10. Junior Kindergarten Math (for Pre-K and junior kindergarten)

11. Learning Chess to Improve math

12. Mom! I Learn Addition Using Math-Chess-Puzzles Connection

13. Mom! I Learn Subtraction Using Math-Chess-Puzzles Connection

14. Mom! I Learn Multiplication Using Math-Chess-Puzzles Connection

15. Mom! I Learn Division Using Math-Chess-Puzzles Connection

16. Frankho ChessDoku 3 by 3

17. Frankho ChessDoku 4 by 4

18. Frankho ChessDoku 5 by 5

介紹何数棋谜

何数棋谜=奧数棋谜 ＋ 思唯腦力開發
英文教材，中英双语教学

只见棋谜不见题　　劝君迷路不哭涕　　数学象棋加谜题　　健脑思维眞神奇

什麼是何数棋谜?

上百篇科學論文已發表國際象棋可以提高兒童問題解答能力. 並且訓練他們的專心及耐力. 所以我們已經知道下國際象棋對兒童有好處. 但是因為國際象棋與計算能力並無直接關係, 所以如何讓兒童能在一個歡樂的環境下也能利用下棋來提高數學的計算呢? 何老師首創並發明有版权的幾何棋藝符號並利用此符號發明了世界第一的独特结合數學与棋谜教材. 何**数棋谜**讓兒童能利用幾何棋藝符號進行邏輯推理及數字的運算. 棋藝與算術的綜合題含蓋了整數, 幾何, 集合, 抽象數, 對比異同, 函數, 座標, 多空間圖形資料, 及規則性數字分析. 並且把棋藝的趣味性和數學的知識性結合在一起.

何**数棋谜**如何幫助兒童腦力思唯的開發?

很簡單的一個道理就是讓學生自願地去用腦, 何**数棋谜**首創獨一無二的融合數學與棋谜的独特趣味寓教於樂教材, 利用國際象棋訓練右腦的座標, 空間分析及圖形處理, 並利用發明了整合棋子與數學的圖形語言, 讓兒童能利用符號圖形訓練左腦進行邏輯推理及數字的運算. 國際象棋與算術的綜合題含蓋了整數, 幾何, 集合, 抽象數, 對比異同, 函數, 多空間圖形資料. 所以枯燥無味的計算題變成了謎題, 學生需要通過更多的思考. 能讓腦去思考愈多則腦力也愈開發. 處里訊息, 分析資料才能發掘出題目. 做這些謎題式數學時可以训練學生比較會專心及有耐心.

何**数棋谜**融合數學與國際象棋的教學理論已在 BC 省數學教師刊物上發表. 科研報告已經證實何**数棋谜**教學法不但可以提高兒童數學解題及思維能力, 還可以開發兒童的腦力, 及分析問題的能力並且增加兒童學習的耐力, 學生的探索創造精神及求知欲. 判斷力, 及自信心等, 啓發思維訓練機警靈巧及加強手腦眼的靈活運用.

Introducing Ho Math Chess™

Ho Math Chess™ = math + puzzles + chess

Frank Ho, a Canadian math teacher, intrigued by the relationships between math and chess after teaching his son chess started **Ho Math Chess™** in 1995. His long-term devotion to research has led his son to become a FIDE chess master and Frank's publications of over 20 math workbooks. Today **Ho Math Chess™** is the world largest and the only franchised scholastic math, chess and puzzles speciality learning centre with worldwide locations. **Ho Math Chess™** is a leading research organization in the field of math, chess, and puzzles integrated teaching methodology.

There are hundreds of articles already published showing chess benefits children and that math puzzles are a very good way of improving brainpower. So, by integrating chess and mathematical chess puzzles together, the learning effect is more significant.

Parents send their children to **Ho Math Chess™** because of they like **Ho Math Chess™** teaching philosophy – offering children problem-solving questions in a variety of formats. The questions could be pure chess, chess puzzles or mathematical chess puzzles in nature of logic, pattern, tree structure, Venn diagram, probability and many more math concepts.

Ho Math Chess™ has developed a series of unique and high-quality math, chess, and puzzles integrated workbooks. **Ho Math Chess™** produced the world's first workbook **Learning Chess to Improve Math.** This workbook is not only for learning chess but also for enriching math ability. This sets **Ho Math Chess** apart from other math learning centres, chess club, or chess classes.

The teaching method at **Ho Math Chess™** is to use math, chess, and puzzles integrated workbooks to teach children fun math. The purposes of **Ho Math Chess™** teaching method and workbooks are to:

- Improve math marks.
- Develop problem-solving and critical thinking skills.
- Improve logic thinking ability.
- Boost brainpower.

Testimonials, sample worksheets, reports, and franchise information can be found at www.homathchess.com.

More information about **Ho Math Chess™** can also be found from the following publications:

1. Why Buy a **Ho Math Chess™** Learning Centre Franchise: A Unique Learning Centre?
2. **Ho Math Chess™** Sudoku Puzzles Sample Worksheets
3. Introduction to **Ho Math Chess™** and its Founder Frank Ho

The above publications can be purchased from www.amazon.com.

www.ingramcontent.com/pod-product-compliance
Lightning Source LLC
Chambersburg PA
CBHW082139210326
41599CB00031B/6031